Numerical Methods for Inverse Problems

To my wife Elisabeth,
to my children David and Jonathan

Series Editor
Nikolaos Limnios

Numerical Methods for Inverse Problems

Michel Kern

iSTE

WILEY

First published 2016 in Great Britain and the United States by ISTE Ltd and John Wiley & Sons, Inc.

ISTE Ltd
27-37 St George's Road
London SW19 4EU
UK

www.iste.co.uk

John Wiley & Sons, Inc.
111 River Street
Hoboken, NJ 07030
USA

www.wiley.com

Library of Congress Control Number: 2016933850

British Library Cataloguing-in-Publication Data
A CIP record for this book is available from the British Library
ISBN 978-1-84821-818-5

Contents

Preface

This book studies methods to concretely address (on a computer) inverse problems. But what is an inverse problem? An inverse problem appears whenever the causes that produced a given effect must be determined, or when we seek to indirectly estimate the parameters of a physical system.

The most common example in the everyday life of many of us comes from the medical field: the medical ultrasound that informs if an unborn baby is in good health involves the solution of an inverse problem. A probe, placed on the belly of the patient, emits and receives ultrasounds. These are deflected, and reflected, by the tissues of the fetus. The sensor receives and interprets these echoes to return an *image* of the contours of these tissues. The image is effectively obtained in an indirect manner. We will see further examples throughout this book.

Intuitively, the observation of an effect may not be sufficient to determine its cause. If I go inside a room and I note that the temperature is (nearly) uniform, it is difficult for me to know what the distribution of temperature was 2 h earlier. It is said that the inverse problem to determine the temperature in the past is "ill-posed". This definition contrasts with the question of determining the future evolution of the temperature, which is, in a sense that we will specify, "well-posed". As Molière's character Monsieur Jourdain does when he speaks prose, it is so common to solve well-posed problems that we (almost) do it without thinking.

Solving inverse problems thus requires the mastery of techniques and specific methods. This book presents some of those chosen for their very general domain of application. It focuses on a small number of methods that will be used in most applications:

– the reformulation of an inverse problem in the form of *minimization of a square error functional*. The reason for this choice is mainly practical: it makes it possible to carry out calculations at a reasonable cost;

– the *regularization* of ill-posed problems and in particular Tikhonov's method;

– the use of the *singular value decomposition* to analyze an ill-posed problem;

– the *adjoint state method* to calculate the gradient of the functionals to minimize when these are not quadratic.

These tools will help to address many (but not all!) inverse problems that arise in practice. Two limitations should be however kept in mind. On the one hand, many inverse problems will make use of different techniques (we will mention a few of them). On the other hand, even when the presented tools can be employed, they are rarely sufficient on their own to completely analyze a complex physical application. Most often, it will be necessary to supplement these tools with a fine analysis of the particular situation to make the most of it (redundancy or not of the data, fast or slow variation of the parameters looked for, etc.).

It is common, in this type of preface, to justify the existence of the presented book! It is true that the question is legitimate (many books already exist on the subject as can be seen in the bibliography), and I do not claim any originality about the content. Nonetheless, readers might still be interested to find a book that discusses both linear and nonlinear problems. In addition, this book can be used as an introduction to the more advanced literature.

This book is aimed at readers with a rather substantial mathematical and a scientific computing background, equivalent to a masters in applied mathematics. Nevertheless, it is a book with a practical perspective. The methods described therein are applicable, and have been applied, and are often illustrated by numerical examples.

The prerequisites to approach this book are unfortunately more numerous that I would have wished. This is a consequence of the fact that the study of inverse problems calls upon many others areas of mathematics. A working knowledge of (both theoretical and numerical) linear algebra is assumed, as is a familiarity with the language of integration theory. Functional analysis, which is what linear algebra becomes when it abandons the finite dimensional setting, is ubiquitous, and the Appendices herein serve as reminders of concepts directly useful in this book. An important part of the examples comes from models of partial differential equations. Here again, the reader will benefit from a prior knowledge of analysis methods (weak formulations, Sobolev spaces) and of numerical analysis (finite element method, discretization schemes for differential equations).

Book layout

We start the book with some general remarks on inverse problems. We will introduce the fundamental concept of *an ill-posed problem*, which is characteristic of inverse problems.

In Chapter 2, we will give several examples of inverse problems, originating from several areas of physics.

An important source of linear inverse problems will be introduced in Chapter 3: the integral equations of the first kind. After outlining the main properties of integral operators, we will show that they lead to ill-posed problems. Finally, we will introduce discretization methods, leading to least squares problems.

The study of these problems is the subject of the subsequent two chapters. In Chapter 4, we will study their mathematical properties in a Hilbertian context: the geometric aspect, and the relationship with normal equations, as well as the questions of existence and uniqueness of the solutions. We will also introduce the fundamental tool, both for theoretical analysis and for numerical approximation, that is the *singular value decomposition*, first for matrices, then for operators between Hilbert spaces. Reminders regarding the numerical aspects of inverse problems can be found in Appendix 1. Techniques for solving ill-posed problems are the subject of Chapter 5, especially Tikhonov's regularization method and spectral truncation. Tikhonov's method will be first addressed from a variational perspective before bringing clarification with singular value decomposition. We will discuss the question of the choice of the regularization parameter and will finish by a short introduction to iterative method.

In the second part, we will discuss nonlinear problems, which are essentially problems of parameters estimation in differential or partial differential equations. In Chapter 6, we will see how to formulate identification problems in terms of minimization and explore the main difficulties that we can expect therefrom. Appendix 2 contains reminders about the basic numerical methods in optimization. Chapter 7 will address the important technique of the adjoint state to compute the functional gradient involved in least squares problems. We will see in several examples how to conduct this computation in an efficient way.

We conclude this second part by briefly introducing issues that could not be discussed in this book, giving some bibliographic hints.

We have compiled reminders regarding the numerical methods of linear algebra for least squares problems, reminders on optimization, as well as some functional analysis results and supplements on linear operators in the appendices.

Acknowledgments

My thanks go first to Professor Limnios, who suggested I write this book, from a first version of course notes that I had published on the Internet. I am grateful to him for giving me the opportunity to publish this course by providing more visibility thereto.

The contents of this book owe a lot, and this is a euphemism, to Guy Chavent. This book grew out of lecture notes that I had written for a course that had originally been taught by G. Chavent, and for which he trusted me enough to let me replace him. Guy was also my thesis supervisor and was the leader of the Inria team where I did all my career. He has been, and remains, a source of inspiration with regard to how to address a scientific problem.

I had the chance to work in the particularly stimulating environment of Inria and to meet colleagues who added great scientific qualities to endearing personalities. I am thinking especially of Jérôme Jaffré and Jean Roberts. A special mention for my colleagues in the Serena team: Hend Benameur, Nathalie Bonte, François Clément, Caroline Japhet, Vincent Martin, Martin Vohralík and Pierre Weis. Thank you for your friendship, and thank you for making our work environment a pleasant and an intellectually stimulating one.

I would like to thank all the colleagues who have told me of errors they found in previous versions of the book, the students of the Pôle Universitaire Léonard de Vinci, of Mines–ParisTech and of the École Nationale d'Ingénieurs of Tunis for listening to me and for their questions, as well as the staff of ISTE publishing for their help in seeing the book through to completion.

Michel KERN
February 2016

PART 1

Introduction and Examples

1

Overview of Inverse Problems

1.1. Direct and inverse problems

According to Keller [KEL 76], two problems are said to be the *inverse* of one another if the formulation of one of them involves the other. This definition includes a degree of arbitrariness and confers a symmetric role to both problems under consideration. A more operational definition is that an inverse problem consists of determining the *causes* knowing the *effects*. Thus, this problem is the inverse of what is called a "direct problem", consisting of the deduction of the effects, the causes being known.

This second definition shows that it is more usual to study *direct* problems. As a matter of fact, since Newton, the notion of causality is rooted in our scientific subconscious, and at a more prosaic level, we have learned to pose, and then solve, problems for which the causes are given, where the objective is to find the effects. This definition also shows that inverse problems may give rise to particular difficulties. We will see further that it is possible to attribute a mathematical content to the sentence "the same causes produce the same effects"; in other words, it is reasonable to require that the direct problem is *well-posed*. On the other hand, it is easy to imagine, and we will see numerous examples, that the same effects may originate from different causes. At the origin, this idea contains the main difficulty of the study of inverse problems: they can have several solutions and it is important to have additional information in order to discriminate between them.

The prediction of the future state of a physical system, knowing its current state, is the typical example of a direct problem. We can consider various inverse problems: for example to reconstitute the past state of the system knowing its current state (if this system is irreversible), or the determination of parameters of the system, knowing (part of) its evolution. This latter problem is that of *the identification* of *parameters*, which will be our main concern in the following.

A practical challenge of the study of inverse problems is that it often requires a good knowledge of the direct problem, which is reflected in the use of a large variety of both physical and mathematical concepts. The success in solving an inverse problem is based, in general, on elements specific to this problem. However, some techniques present an extended application domain and this book is an introduction to the principal techniques: the regularization of ill-posed problems and the least squares method.

The most important technique is the reformulation of an inverse problem in the form of the minimization of an error functional between the actual measurements and the synthetic measurements (that is the solution to the direct problem). It will be convenient to distinguish between linear and nonlinear problems. It should be noted here that the nonlinearity in question refers to the inverse problem, and that the direct problem itself may or may not be linear.

In the case of linear problems, resorting to linear algebra and to functional analysis allows accurate results as well as efficient algorithms to be obtained. The fundamental tool here is the singular value decomposition of the operator, or of the matrix, being considered. We will study the regularization method in detail, which consists of slightly "modifying" the problem under study by another that has "better" properties. This will be specified in Chapters 4 and 5.

Nonlinear problems are more difficult and there exist less overall results. We will study the application of optimization algorithms to problems obtained by the reformulation referred to above. A crucial technical ingredient (from the numerical point of view) is the calculation of the gradient of the functional to be minimized. We will study the adjoint state method in Chapter 7. It allows this calculation at a cost that is a (small) multiple of that of solving the direct problem.

As it can be seen, the content of this book primarily aims to present *numerical methods* to address inverse problems. This does not mean that *theoretical* questions do not exist, or are devoid of interest. The deliberate choice of not addressing them is dictated by the practical orientation of the course, by the author's taste and knowledge, but also by the high mathematical level that these issues require.

1.2. Well-posed and ill-posed problems

In his famous book, Hadamard [HAD 23] introduced, as early as 1932, the notion of a *well-posed problem*. It concerns a problem for which:

– a solution exists;

– the solution is unique;

– the solution depends continuously on the data.

Of course, these concepts must be clarified by the choice of space (and of topologies) to which the data and the solution belong.

In the same book, Hadamard suggested (and it was a widespread opinion until recently) that only a well-posed problem could properly model a physical phenomenon. After all, these three conditions seem very natural. In fact, we shall see that inverse problems often do not satisfy either of these conditions, or even the three all together. Upon reflection, this is not so surprising:

– a physical model being established, the experimental data available are generally noisy and there is no guarantee that such data originate from this model, even for another set of parameters;

– if a solution exists, it is perfectly conceivable (and we will see examples of this) that different parameters may result in the same observations.

The absence of one or any other of the three Hadamard's conditions does not have the same importance with respect to being able to solve (in a sense that remains to be defined) the associated problem:

– the fact that the solution of an inverse problem may not exist is not a serious difficulty. It is usually possible to restore the existence by relaxing the concept of solution (a classic procedure in mathematics);

– the non-uniqueness is a more serious problem. If a problem has several solutions, there should be a means of distinguishing between them. This requires additional information (we speak of *a priori* information);

– the lack of continuity is probably the most problematic, in particular in view of an approximate or a numerical solution. Lack of continuity means that it is not possible (regardless of the numerical method) to approach a satisfactory solution of the inverse problem, since the data available will be noisy, therefore close to the actual data, but different from the actual data.

A problem that is not well-posed within the meaning of the definition above is said to be *ill-posed*. We now give an example that, although very simple, illustrates the difficulties that may be found in more general situations.

EXAMPLE 1.1.– Differentiation and integration are two problems that are the inverse of each other. It would seem more natural to consider differentiation as the direct problem and integration as the inverse problem. In fact, integration has good mathematical properties that lead to consider it as the direct problem. In addition, differentiation is the *prototypical* ill-posed problem, as we shall see in the following.

Consider the Hilbert space $L^2(\Omega)$, and the integral operator A defined by

$$Af(x) = \int_0^x f(t)\, dt. \qquad [1.1]$$

It is easy to directly see that $A \in \mathcal{L}(L^2(0,1))$, or theorem 3.1, can be applied (see example 3.1). This operator is injective; however, its image is the vector subspace

$$\operatorname{Im} A = \{f \in H^1(0,1),\ u(0) = 0\}$$

where $H^1(0,1)$ is the Sobolev space. In effect, the equation

$$Af = g$$

is equivalent to

$$f(x) = g'(x) \text{ and } g(0) = 0.$$

The image of A is *not* closed in $L^2(0,1)$ (of course, it is closed in $H^1(0,1)$). As a result, the inverse of A is not continuous on $L^2(0,1)$, as shown in the following example.

Consider a function $f \in C^1([0,1])$, and let $n \in \mathbf{N}$. Let

$$f_n(x) = f(x) + \frac{1}{n}\sin\left(n^2 x\right),$$

then

$$f_n'(x) = f'(x) + n\cos\left(n^2 x\right).$$

Simple calculations show that

$$\|f - f_n\|_2 = \frac{1}{n}\left(\frac{1}{2} - \frac{1}{4n}\sin\left(2n^2\right)\right)^{1/2} = 0\left(\frac{1}{n}\right),$$

whereas

$$\|f' - f_n'\|_2 = n\left(\frac{1}{2} + \frac{1}{4n}\sin\left(2n^2\right)\right)^{1/2} = O(n).$$

Thus, the difference between f' and f'_n may be arbitrarily large, even though the difference between f and f' is arbitrarily small. The derivation operator (the inverse of A) is thus not continuous, at least with this choice of norms.

The instability of the inverse is typical of ill-posed problems. A small perturbation over the data (here f) can have an arbitrarily large influence on the result (here f').

A second class of inverse problems is the estimation of parameters in differential equations. We are going to discuss a very simple example of this situation.

EXAMPLE 1.2.– Considering the elliptic problem in one dimension:

$$\begin{cases} -(a(x)u'(x))' = f(x), & \text{for } x \in]-1,1[\\ u(-1) = u(1) = 0. \end{cases} \qquad [1.2]$$

This equation, or other similar although more complex, arises in several examples in the following chapter. In this example, we choose $a(x) = x^2 + 1$, and the solution $u(x) = (1 - x^2)/2$, which gives $f(x) = 3x^2 + 1$.

The direct problem consists of calculating u, given a and f. For the inverse problem, we shall consider that f is known, and we will try to recover the coefficient a from a measurement of u. For this example, voluntarily simplified, we shall assume that u is measured over the whole interval $]-1,1[$, which is obviously unrealistic. We shall see that even in this optimistic situation, we are likely to face difficulties.

By integrating equation [1.2], and by dividing by u', we obtain the following expression for a (assuming that u' does not vanish, which is not true in our example):

$$a(x) = \frac{C}{u'(x)} + \frac{1}{u'(x)} \int_0^x f(\xi)\, d\xi, \qquad [1.3]$$

which yields in our particular case:

$$a(x) = \frac{C}{x} + x^2 + 1 \quad \text{for } x \neq 0, \qquad [1.4]$$

where C is an integration constant.

We can see that even in this particular case, a is not determined by the data, that is u. Of course in this case, it is clear that the "correct" solution corresponds to $C = 0$, since this is the only value for which a is bounded. In order to be able to discriminate among the various possible solutions, we resort to additional information (usually referred to as *a priori information*).

In this problem, there are two sources of instability: first, equation [1.3] involves u', and we just have seen that the transition from u to u' causes instability. This is a phenomenon common to linear and nonlinear problems. On the other hand, the division by u' shows an instability specific to nonlinear problems. If u' vanishes at some point, the division is impossible. If u' is simply small, the division will be a cause of instability.

This book is dedicated to the study of methods allowing for the recovery of a certain degree of stability in ill-posed problems. It is however necessary to keep in mind this observation from [ENG 96]: "no mathematical trick can make an inherently unstable problem stable". The methods we are going to introduce in the following will make the problem under consideration stable, but at the price of a modification of the solved problem (and therefore of its solution).

2

Examples of Inverse Problems

In this chapter, we present a few "concrete" examples of inverse problems, as they occur in the sciences or in engineering. This list is far from exhaustive (see the references at the end of this chapter for other applications).

Among the areas in which inverse problems play an important role, we can mention the following:

– medical imaging (ultrasound, scanners, X-rays, etc.);

– petroleum engineering (seismic prospection, magnetic methods, identification of the permeabilities in a reservoir etc.);

– hydrogeology (identification of the hydraulic permeabilities);

– chemistry (determination of reaction constants);

– radars (determination of the shape of an obstacle);

– underwater acoustics (same objective);

– quantum mechanics (determination of the potential);

– image processing (restoration of blurred images).

From a mathematical point of view, these problems are divided into two major groups:

– linear problems (echography, image processing, etc.), which amount to solving an integral equation of the first kind;

– nonlinear problems, which are mostly questions of parameter estimation in differential or partial differential equations.

2.1. Inverse problems in heat transfer

In order to determine the temperature distribution in an inhomogeneous material occupying a domain (open connected subset) Ω of $\mathbf{R}^3$, the conservation of energy is first written as

$$\rho c \frac{\partial T}{\partial t} + \operatorname{div}(\vec{q}) = f(x, y, z) \quad \text{in } \Omega \qquad [2.1]$$

where T is the temperature, ρ is the density of the fluid, c is the specific heat, $\vec{q}$ represents a heat flux and f is a volume source.

Fourier's law then connects the heat flux density to the temperature gradient:

$$\vec{q} = -K \operatorname{grad} T, \qquad [2.2]$$

where K is the thermal conductivity (which may be a tensor, and depends on the position).

By eliminating $\vec{q}$, we obtain the equation for the temperature, known as the heat equation, in a heterogeneous medium:

$$\rho c \frac{\partial T}{\partial t} - \operatorname{div}(K \operatorname{grad} T) = f \quad \text{in } \Omega. \qquad [2.3]$$

This equation must be complemented by boundary conditions on the boundary of the domain Ω and an initial condition.

The direct problem is to determine T knowing the physical coefficients ρ, c and K as well as the source of heat f. This problem is well known, both from the theoretical point of view (existence and uniqueness of the solution) and the numerical point of view. Several inverse problems can be established:

– given a measurement of the temperature at an instant $t_f > 0$, determine the initial temperature. We will discuss it in example 2.1;

– given a (partial) temperature measurement, determine some of the coefficients of the equation.

Note that the first of these problems is linear, while the second is nonlinear: in fact, the application $(\rho, c, K) \mapsto T$ is nonlinear.

EXAMPLE 2.1 (Backward heat equation).– We consider the ideal case of a homogeneous and infinite material (in one spatial dimension to simplify). The temperature is a solution of the *heat equation*:

$$\frac{\partial T}{\partial t} - \frac{\partial^2 T}{\partial x^2} = 0 \qquad [2.4]$$

(there is no source). It is assumed that the temperature is known at some time t_f, or $T_f(x) = T(x, t_f)$, and that the objective is to find the initial temperature $T_0(x) = T(x, 0)$.

The problem of determining T_f knowing T_0 is the Cauchy problem for the heat equation. It has a unique solution, which continuously depends on the initial data. As we shall see, this is not true for the inverse problem that we consider here. Physically, this is due to the irreversible character of the thermal diffusion. It is well known that the temperature tends to become homogenized over time, and this implies that it is not possible to go back, that is to recover the previous state that can be more heterogeneous than the current state.

Because of the very simplified situation that we have chosen, we can calculate by hand the solution of the heat equation [2.4]. Using the spatial Fourier transform of equation [2.4] (we note $\hat{T}(k, t)$ the Fourier transform of $T(x, t)$ keeping t as fixed), we obtain an ordinary differential equation (where this time it is k that is used as a parameter) whose solution is

$$\hat{T}_f(k) = e^{-|k|^2 t_f} \hat{T}_0(k). \qquad [2.5]$$

Using the inverse Fourier transform, we can see that the solution at the instant t_f is related to the initial condition by a convolution with the elementary solution of the heat equation:

$$T_f(x) = \frac{1}{2\sqrt{\pi t_f}} \int_{-\infty}^{+\infty} e^{-(x-y)^2/4t_f}\, T_0(y)\, dy. \qquad [2.6]$$

It is well known [CAN 84] that, for any "reasonable" function T_0 (continuous, bounded), the function T_f is infinitely differentiable, which mathematically expresses the irreversibility mentioned earlier.

While remaining in the Fourier domain, we can pointwise invert equation [2.5], but the function

$$k \mapsto e^{|k|^2 t}\, \hat{T}_f(k)$$

will be in $L^2(\mathbf{R})$ only for functions T_f decreasing very quickly at infinity, which is a very severe restriction. A temperature measured experimentally has little chance of satisfying this condition and this is what causes the instability of the inverse problem.

We will meet again the analogue of this condition in Chapter 4 (section 4.4.2) when we will study the Picard condition.

We shall now continue with some examples of parameter estimation.

EXAMPLE 2.2 (Identification of the diffusion coefficient in a steady-state thermal model).– In order to simplify, we consider only the steady state and we assume that the boundary of the domain is maintained at a temperature of 0. The heat equation [2.3] and the boundary condition then give:

$$\begin{cases} -\operatorname{div}(K \operatorname{grad} T) = f(x,y,z) & \text{in } \Omega \\ T = 0 & \text{on } \partial\Omega \end{cases} \qquad [2.7]$$

The direct problem assumes as known the thermal conductivity K and the heat source f and aims to determine the distribution of the temperature T at every point in the material. This is the prototype of an elliptic equation of the second order and its properties are well known: under reasonable hypotheses about K ($K \in L^\infty(\Omega)$ with $0 < K_* \leq K \leq K^* < \infty$) and about f ($f \in L^2(\Omega)$, it admits a unique solution, according to the Lax–Milgram theorem. In addition, a numerical calculation with a finite element method is all the more standard (see [DAU 90, Chapter V]).

The same cannot be said about the inverse problem. In order to specify it, it is first of all necessary to clarify which *measurements* are available. This depends obviously on the experimental device being used, but in any case, it will typically be unrealistic to assume that the temperature is known erywhere. In our case, these observations could be, for example, measurements of the temperature at several locations inside the material, or heat flux measurements $-k\dfrac{\partial T}{\partial n}$ on the boundary of the domain (which is in this case referred to as boundary observation). The inverse problem consists then of searching the (or a) conductivity function, such that there is a function T solution of [2.7], which coincides with the observations.

We can immediately observe several possible difficulties as follows:

– first, in obtaining the observations. An experiment is never easy to carry out. In our example, it is not realistic to assume that the temperature can be measured at all points of the domain (think of a room in an apartment);

– moreover, there is a risk that not enough observations be available relative to the number of parameters that are being sought. Here, if only a boundary observation is available, that is of a two-variable function, it will be difficult to find a three-variable function, regardless of the method being used;

– in particular, we can immediately see that if the temperature is constant in a subdomain of Ω, the conductivity within that subdomain is not determined. Therefore, additional information should be available to fill this lack of measurement;

– finally, any measurement contains errors and besides the mathematical model [2.7] does not accurately reflect reality. Thus, there is actually no reason for the inverse problem to have a solution.

This difficult problem has been the subject of numerous studies, both theoretical and numerical (see [ISA 98] for an introduction). It should be noted that the same model is involved in other areas of application (medical, geophysical prospection by electrical or magnetic methods, etc.)

EXAMPLE 2.3 (Identification in one spatial dimension).– We can better understand this example by reducing it to a single dimension. The goal is to determine the function $K(x)$ from the equation

$$-(K(x)T'(x))' = f(x), \quad \text{for } x \in]0,1[\qquad [2.8]$$

(with the appropriate boundary conditions), and knowledge of T. In order to simplify, we will assume that we know T at all points of the interval $]0,1[$. Under the assumption that T' does not vanish at any point of the interval (which is not necessarily satisfied), we can integrate equation [2.8] to obtain

$$K(x) = \frac{1}{T'(x)}\left(K(0)T'(0) - \int_0^x f(t)\,dt\right). \qquad [2.9]$$

This equation is the same as the one we had obtained in example 1.2. We had seen therein the two possible instability sources: the first related to the appearance of a derivative of T, the other related to the division by T'. This second effect is essentially nonlinear. If $T'(x)$ is small, the division by $T'(x)$ is likely to magnify the errors already present in the measurement of T. This is obviously related to the fact that if T' vanishes, K is not determined at all.

For more information about this example, the reader may consult the article by Engl [ENG 93] or the book by Engl *et al.* [ENG 96].

2.2. Inverse problems in hydrogeology

Hydrogeology, or the study of groundwater, is another abundant source of inverse problems. In effect, it is difficult to access the underground layers to measure the aqueous properties of rocks. A currently topical problem is the control of pollutants in groundwater. To mention only one practical example, Mosé's habilitation thesis [MOS 98] studied the influence on water quality of the accident of a truck that was carrying CCl4 gas (carbon tetrachloride) in the East of France in 1970 (see also [VIG 83]). An essential parameter of this study is the hydraulic conductivity of the subsurface, which obviously depends on the position.

There are a wide variety of physical models, including various approximations. We present one such situation below, based on Siegel's thesis [SIE 95] (also see [BEA 87, MAR 86, SUN 94]):

EXAMPLE 2.4 (Transport of a pollutant by an aquifer).– A porous medium consists of a rocky matrix, comprising pores that allow water to pass through. It is essentially impossible to describe the flow of a fluid in such a heterogeneous medium, insofar as we must take into account the spatial scales ranging from the centimeter (pore) to the kilometer (the regional model), and that the accurate disposition of the pores is not known in any case. Simplified physical models are then used, the most common being Darcy's law, which relates the height of water in the medium, called *piezometric head* and denoted $h(x, y, z, t)$, to the filtration velocity $\vec{q}(x, y, z, t)$. This law expresses that the velocity is proportional to the opposite of the hydraulic gradient:

$$\vec{q} = -K \operatorname{grad} h \qquad [2.10]$$

where K is the coefficient of hydraulic conductivity. This may in principle be a tensor but we will limit ourselves to the case in which it is a scalar.

The conservation of mass is also expressed (under the assumption that the medium is incompressible):

$$S\frac{\partial h}{\partial t} + \operatorname{div} \vec{q} = f \qquad [2.11]$$

where S is the coefficient of specific storage and f is a source (assumed as known).

The elimination of $\vec{q}$ yields for h the parabolic equation:

$$S\frac{\partial h}{\partial t} - \operatorname{div}\left(K \operatorname{grad} h\right) = f \qquad [2.12]$$

to which initial conditions are added (h given at $t = 0$) as well as boundary conditions (Dirichlet, corresponding to an imposed head, or Neumann, corresponding to an imposed flux).

In addition to the flow, contaminant transport issues involve the way in which the concentration of a species (chemical compound, hydrocarbon, radionuclide) transported by the flow evolve. This phenomenon brings forward three mechanisms: convection (imposed by the filtration velocity $\vec{q}$), molecular diffusion and kinematics dispersion. We will not describe these last two mechanisms in detail (to this end, see the references cited previously). The studied quantity is the concentration $C(x, y, z, t)$ of the pollutant, which obeys a convection–diffusion type equation:

$$\varepsilon \frac{\partial C}{\partial t} + \operatorname{div}(\vec{q}C) - \operatorname{div}(D \operatorname{grad} C) = f_c \qquad [2.13]$$

where ε is the porosity (fraction of pores occupied by moving water), D is the diffusion tensor (aggregating the molecular diffusion and the kinematics dispersion) and f_c is a potential source of pollutant. An initial condition is added (known concentration at the initial instant) as well as boundary conditions.

The direct problem is constituted of equations [2.12] and [2.13]. This coupled problem is theoretically nonlinear, due to the term $\operatorname{div}(\vec{q}C)$. In practice, however, it is often possible to solve first equation [2.12], then [2.13], $\vec{q}$ being known.

As an example, assume that the concentration is measured at a certain number of points and at discrete instants (it is not realistic here to assume that the measurement is continuous in time). We thus know $C(x_o, y_o, z_o, t_o)$, $o = 1, \ldots, N_o$. The inverse problem is then to find the hydraulic conductivity (and to a lesser extent the other parameters of the model), knowing these measurements. This problem is underdetermined, because it is unusual that sufficient measurements are available.

EXAMPLE 2.5 (The steady-state case).– Consider the steady state in the above model. In this case, only equation [2.12] is taken into account, where it is assumed that $S = 0$ and that the source f is time independent. It then simply yields an elliptic equation of the second order: $-\operatorname{div}(K \operatorname{grad} h) = f$ in a domain Ω. Taking, for example, the case where the piezoelectric head is imposed on the boundary $\partial\Omega$. The flux $K\frac{\partial h}{\partial n}$ is measured on the boundary and the aim is as always to identify the coefficient K. Here again, we find an elliptic model of the type of the one studied in section 2.1.

EXAMPLE 2.6 (One-dimensional hydrogeology).– Here again, we shall consider the simplified problem where the flow is essentially 1D in one horizontal direction that we take as axis Ox. Such a model is obtained by integrating over vertical layers of the previous model. The equations are written:

$$\begin{cases} S\dfrac{\partial h}{\partial t} - \dfrac{\partial}{\partial x}\left(K\dfrac{\partial h}{\partial x}\right) = f & \text{in } [0,L] \times]0,T[\,, \\ q = -K\dfrac{\partial h}{\partial x} & \text{in } [0,L] \times]0,T[\,, \\ \varepsilon\dfrac{\partial C}{\partial t} + \dfrac{\partial(\vec{q}C)}{\partial z} - \dfrac{\partial}{\partial x}\left(D\dfrac{\partial C}{\partial x}\right) = f_c & \text{in } [0,L] \times]0,T[\,. \end{cases} \qquad [2.14]$$

with (for example) given initial conditions, h defined at the two ends and C given at $x = 0$. A measurement of $C(L,t)$ is given and we aim to identify $K(x)$.

2.3. Inverse problems in seismic exploration

Oil exploration by seismic methods (and seismology) gives rise to an inverse problem that has been widely studied because of the economical benefits that are incurred from its solution. It is actually a family of inverse problems, whose common goal is to determine the elastic properties of the subsurface (density, propagation speed of the elastic waves) from measurements of displacement or pressure fields on the surface.

During a seismic campaign (see Figure 2.1), a source (typically, an explosive device) causes a disturbance in the rocks forming the subsurface. The echoes are recorded by a series of sensors placed on the surface. This experience is repeated for several positions of the source (from several hundreds to several thousands). In this way, a very large amount of data is measured (that can reach hundreds of gigabyte). The goal is, once again, to estimate the properties of the medium given a propagation model. The geophysical community has developed a large amount of specific methods to address this problem. The book [BLE 00] presents these methods in a synthetic way.

EXAMPLE 2.7 (The acoustic model).– There are several physical models that can account (with varying degrees of approximation) for the experiment described above. We shall confine ourselves to study one of the simplest: we make the hypothesis that the region under study consists of a fluid (which corresponds to a marine seismic experiment). In this case, it can be demonstrated (see [DAU 90, vol. 1]) that the wave propagation is governed by *the acoustic wave equation*, and the measured quantity is a pressure (scalar) field. It is convenient to assume that the domain of study is the half-space $\{z > 0\}$ (the Earth is obviously neither flat nor infinite but these approximations are justified by the scales being considered, which are here in the order of a few kilometers), the axis Oz being oriented downward. We will denote by

$\Omega = \mathbf{R}^2 \times \mathbf{R}_+$ the spatial domain of study and T the duration of the experiment, as well as

$p = p(x, y, z, t)$	the pressure,
$\rho = \rho(x, y, z)$	the density,
$c = c(x, y, z)$	the propagation speed,
$f = f(x, y, z, t)$	the source.

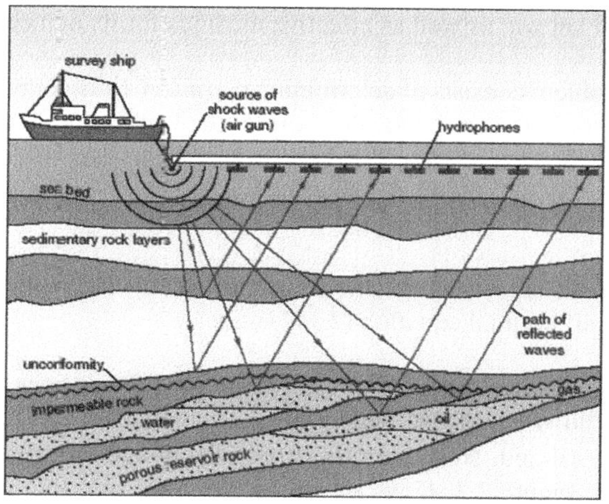

Figure 2.1. *Acquisition device in marine seismology. Source http://www.subsurfwiki.org/. For a color version of the figure, see www.iste.co.uk/kern/inverse.zip*

The direct problem then amounts, knowing f, c and ρ to finding p, a solution of

$$\begin{cases} \dfrac{1}{\rho c^2}\dfrac{\partial^2 p}{\partial t^2} - \operatorname{div}\left(\dfrac{1}{\rho}\operatorname{grad} p\right) = f & \text{in } \Omega \times]0, T[, \\ p(x, y, z, 0) = \dfrac{\partial p}{\partial t}(x, y, z, 0) = 0 & \text{in } \Omega, \\ \dfrac{\partial p}{\partial z}(x, y, 0, t) = 0 & \text{on } \{z = 0\}. \end{cases} \qquad [2.15]$$

The first equation of [2.15] represents Newton's law. The second is the initial state of the system (assumed here to be at rest), the third is a boundary condition, here expressing that there is no stress on the surface of the ground. It is well known (see, for example, Dautray-Lions [DAU 90, vol. 5]) that under the reasonable hypotheses on the coefficients ρ and c and the source f, [2.15] admits a unique solution p, which depends continuously from f. It is also true, but more difficult to prove because the dependence is nonlinear that p depends continuously on c and ρ.

It should be noted that equation [2.15] also describes the motion of the SH waves in an elastic solid, under the assumption that the Earth is two dimensional (the parameters c, ρ and p then have different physical meanings).

Once more, this direct problem has been extensively studied, its numerical properties are well known, as well as effective methods for its numerical solution.

The inverse problem consists of determining c, ρ and f (it is not realistic to assume that f is known) from measurements, that is from the knowledge of the set

$$\{p(x_g, y_g, z_g, t),\ g = 1, \ldots, N_g,\ t \in [0, T]\}.$$

This problem is nonlinear, since the solution p depends nonlinearly on c and ρ, even if the partial differential equation [2.15] is linear.

Figure 2.2 presents a synthetic, but realistic model. It has been proposed by the Insttut Français du Pétrole in 1988, in the context of a "blind" inversion experiment [VER 90] and, since then, is part of the commonly used models to test inversion methods. Figure 2.3 shows an example of a seismic shot performed on the Marmousi model. It shows a set of traces, each trace (a function of time represented vertically) corresponds to the recording of a geophone. The traces are juxtaposed so that the structures present in the model induce structures in the seismogram.

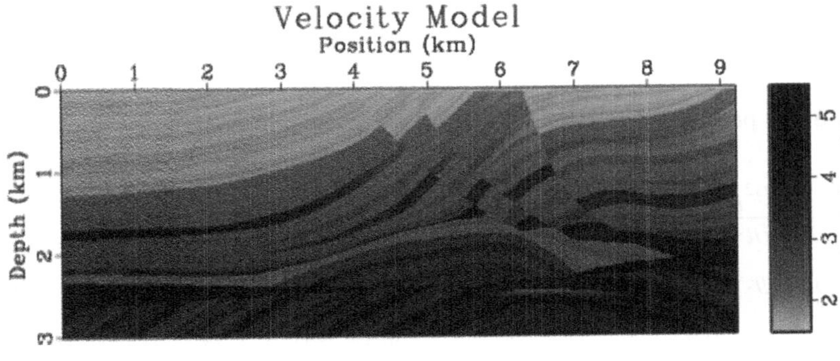

Figure 2.2. *"Marmousi" geological model. Source http://www.reproducibility.org/, Trevor Irons*

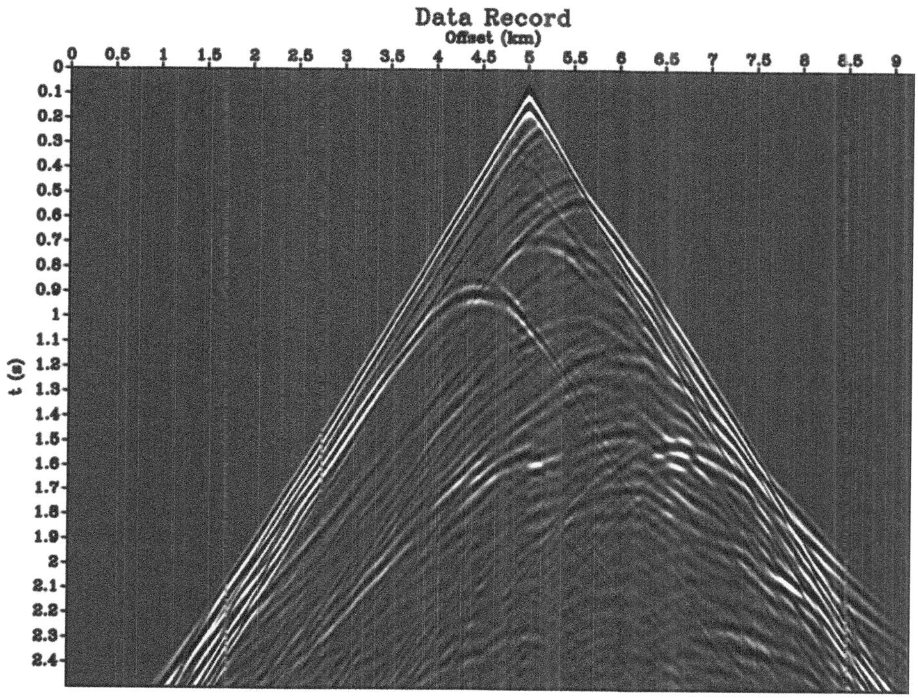

Figure 2.3. *A synthetic shoting on the Marmousi model. Source http://www.reproducibility.org/, Trevor Irons*

It is not realistic to assume that we know the pressure at each instant, at every point of the domain $\Omega\times]0,T[$. Sensors are available (*geophones* in the case of terrestrial seismology, *hydrophones* in that of marine seismology), and to simplify, it will be assumed that records are measured at discrete points, but are continuous in time (this is again justified by the consideration of the scales of phenomena). The number of sensors will be denoted by N_g, and their position by (x_g, y_g, z_g), $g = 1, \ldots, N_g$. The simulated measurements $p(x_g, y_g, z_g, t)$, $g = 1, \ldots, N_g$, $t \in [0, T]$ are then extracted from the pressure field solution of [2.15].

In reality, there is an additional parameter: the experiment described above is repeated by moving the source–receiver device. All of these *shots* provide a huge amount of data. This inverse problem is overdetermined. An important piece of information to take advantage of is that all these records originate from the same subsurface. This observation, seemingly trivial, is the basis of several recent inversion methods [CLÉ 01, SYM 94, SYM 93].

It is commonly assumed that the density of the medium is constant. In this case, equation [2.15] becomes the usual wave equation:

$$\begin{cases} \dfrac{1}{c^2}\dfrac{\partial^2 p}{\partial t^2} - \Delta p = f & \text{in } \Omega \times]0, T[, \\ p(x, y, z, 0) = \dfrac{\partial p}{\partial t}(x, y, z, 0) = 0 & \text{in } \Omega, \\ \dfrac{\partial p}{\partial z}(x, y, 0, t) = 0 & \text{on } \{z = 0\}. \end{cases} \qquad [2.16]$$

REMARK 2.1.– Before considering models even simpler than this one, it should be highlighted in which aspects this model itself is simplified: in reality, the Earth is not described by an acoustic model. A more realistic model should be used that better accounts for the physics. A large number of models are possible, depending upon the phenomena we want to consider: 2D or 3D, acoustic, elastic (the Earth is a solid), viscoelastic (taking into account the damping mechanisms in the soil), isotropic or anisotropic, etc. But in fact, it is not obvious that a refined model will be superior to the model we have just mentioned. As a matter of fact, it is not useful to add parameters to a model if one is not capable of measuring additional data, which would allow them to be determined. Thus, our model may represent a reasonable compromise. A similar discussion (concerning the choice between 2D and 3D models) can be found in the book of Bleistein and *et al.* [BLE 00]. More recently, due to advances in parallel computing, 3D models have become affordable, see [EPA 08, MON 15] or [MÉT 13].

EXAMPLE 2.8 (The stratified model).– The example we have just studied, although already simplified relative to the real situation is still complicated: the direct problem requires the solution of a wave equation for each source position; furthermore, the number of unknowns necessary to represent the velocity can become huge in a complex 3D geology situation. In fact, if 3D modeling is within the reach of modern supercomputers, inversion must for the moment be limited to 2D models and these are extremely expensive. Therefore, approximations of the acoustic model are still being used. One of them, that will be intuitively justified by looking at a rocky mountain landscape, is to assume that the Earth is *stratified*, that is that the parameters ρ and c depend only on the depth (the variable z). In addition, if we assume that the source f is a plane wave (and it is possible by means of an appropriate processing of the data to approximately reach this situation), the pressure p only depends on z (and on time), and equation [2.15] becomes:

$$\begin{cases} \dfrac{1}{\rho c^2}\dfrac{\partial^2 p}{\partial t^2} - \dfrac{\partial}{\partial z}\left(\dfrac{1}{\rho}\dfrac{\partial p}{\partial z}\right) = f & \text{in } [0, Z] \times]0, T[, \\ p(z, 0) = \dfrac{\partial p}{\partial t}(z, 0) = 0 & \text{in } [0, Z], \\ p(0, t) = 0 & \text{over } \{z = 0\}, \end{cases} \qquad [2.17]$$

and $p(z_G, t)$, for $t \in [0, T]$ is measured (the same simplification occurs in equation [2.16]).

This problem is simpler, since equation [2.17] is in 1D space. For this problem, a rather large number of results are known (see, for example, [BAM 77, BAM 79, BAM 82]). This example already shows the essential difficulties of the general case but allows for an economical solution of the direct problem.

We have just seen that inverse problems can be devised for the three common types of partial differential equations: hyperbolic, parabolic and elliptic. We consider now examples of a different type, leading to integral equations.

2.4. Medical imaging

Medical sciences provide a large number of inverse problems, whose practical importance is obvious. We are going to quickly discuss a few of them. The description that we present here is borrowed from the article by Louis [LOU 92], where more details can be found.

EXAMPLE 2.9 (X-ray tomography).– This is the technique used by scanners. An X-ray tube is mounted on a gantry that surrounds the patient. The emitted rays are measured by sensors placed in the front of the transmitter. We consider the 2D situation, where the domain represents a transversal section of the patient. It is assumed that the rays follow a straight line and are attenuated when traversing the tissues, proportionally to the intensity itself and to the travelled distance (Bouger's law). X-rays follow straight lines and we shall parameterize these lines by their normal vector $u \in \mathbf{R}^2$, and their distance s to the origin (see Figure 2.4).

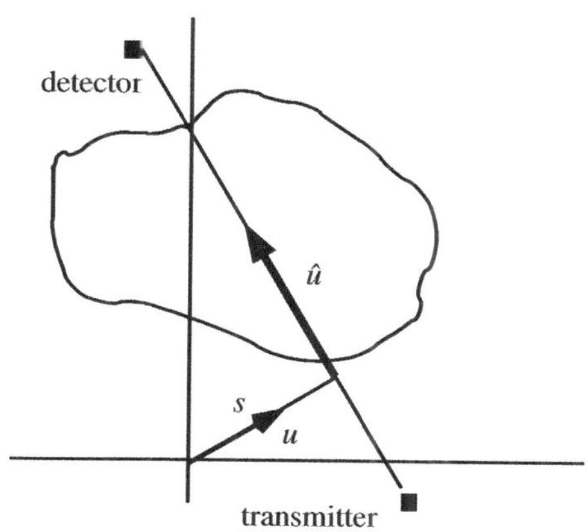

Figure 2.4. *Geometry of the experiment*

We denote the attenuation coefficient by f (which may depend on position), and we obtain the following equation for the intensity of the ray at a distance s from the origin, and at a length t along the line:

$$\Delta I(su + t\hat{u}) = -I(su + t\hat{u})f(su + t\hat{u})\Delta t,$$

where $\hat{u}$ is a unit vector orthogonal to u. By letting Δt tend to 0, the following differential equation is obtained:

$$\frac{d}{dt}I(su + t\hat{u}) = -I(su + t\hat{u})f(su + t\hat{u}).$$

By denoting I_0 and I_L as the intensities to the transmitter and the receiver, respectively (we will assume they are both outside the object, which amounts to state that there are at infinity), the previous differential equation is integrated as

$$-\ln \frac{I_l(s,u)}{I_0(s,u)} = \int_{\mathbf{R}} f(su + t\hat{u})\, dt. \qquad [2.18]$$

The direct problem consists of determining the measured intensity at the detector by knowing the intensity at the transmitter as well as the attenuation function f. The inverse problem is therefore to determine the function f knowing the two intensities.

The integral operator involved in the right-hand side of the previous equation is called the *Radon transform* of f, after the Austrian mathematician J. Radon, who also provided (in 1917) the inversion formula allowing in principle the reconstruction of the attenuation function f from the knowledge of the transforms on all the lines of the plane. The expression *in principle* in the previous sentence means that the Radon transform is known exactly, and thus represents an important caveat. In effect, the inversion formula assumes that Rf is known for all directions u. This means, in practice, that the data must be measured in a roughly uniform manner on a circle around the patient (which may or may not be feasible). If this is not the case, the problem is much trickier and it is difficult to recover f in a stable way. Moreover, as we shall see in a special case below, the reconstruction formula involves the derivative of the measurements, which also shows its unstable character. This problem is discussed in detail in [HER 80, NAT 86].

We will once more assume that we are addressing a simplified situation, where calculations are accessible.

EXAMPLE 2.10 (Tomography with circular symmetry).– We now assume that the medium is a circle of radius ρ and that the function f depends only on the distance s

$(f(s,u) = F(s))$. In addition, all of the information is contained in the integral taken according to a single line, whose direction is denoted u_0. If we let

$$g(s) = -\ln \frac{I_t(s,u_0)}{I_0(s,u_0)},$$

equation [2.18] becomes:

$$\int_0^\rho \frac{rF(r)}{\sqrt{r^2 - s^2}}\, dr = \frac{g(s)}{2}. \qquad [2.19]$$

This is an Abel integral equation, which is used in several applications (we will return to it in example 2.13). A comprehensive reference about this equation is given in [GOR 91]. It is shown that equation [2.19] has a (unique) solution given by

$$F(r) = -\frac{1}{\pi} \int_r^\rho \frac{g'(s)}{\sqrt{s^2 - r^2}}\, ds. \qquad [2.20]$$

Once more, this formula makes use of the derivative of the data g.

EXAMPLE 2.11 (Echography).– This investigation method presents the great advantage of being without risks to the patient. The sources here are brief pulses of a high-frequency acoustic wave; the measurements are acoustic echoes and the discontinuities of the propagation speed of the medium must be found. The direct problem is to calculate u_s knowing q (and u_i) and the inverse problem is to find q from measurements of u_s performed at a distance from the obstacle. Relatively to the examples of section 2.3, the problem is defined here in the frequency domain. When the wave passes through the patient, it is reflected by changes in density and the elastic parameters of the medium.

We denote by ω the frequency of the source, ρ the density and p the pressure. Letting $u = \rho^{-1/2}p$, we obtain the Helmholtz equation:

$$\Delta u + (k^2 + q)u = 0 \qquad [2.21]$$

where $k = \omega/c$ and the potential q is related to the density by

$$q = \frac{1}{2\rho}\Delta\rho - \frac{3}{4}\left(\frac{1}{\rho}\nabla\rho\right)^2.$$

We consider that the source is a plane wave, denoted by u_i, which is a solution of the equation without the obstacle

$$(\Delta + k^2)u_i = 0,$$

and we define the diffracted wave u_s (s for *scattered*) by $u_s = u - u_i$. The diffracted wave is then a solution of

$$(\Delta u + k^2)u = -q(u_i + u_s) \qquad [2.22]$$

The Green function for the operator $\Delta + k^2$ is then introduced

$$G(x,y) = \begin{cases} \dfrac{i}{4} H_0^{(1)}(k|x-y|) & \text{in two dimensions,} \\ \dfrac{1}{4\pi} e^{ik|x-y|}/|x-y| & \text{in three dimension ,} \end{cases}$$

where $H_0^{(1)}$ is the Hankel function of the first kind with order 0. It can then be shown that equation [2.22] can be presented in the form of an integral equation (known as the Lippmann–Schwinger equation):

$$u_s(x) = k^2 \int_\Omega G(x,y)q(y)(u_i + u_s)(y)\, dy. \qquad [2.23]$$

Because of the presence of the term qu_s on the right-hand side, the inverse problem is nonlinear. A reasonable approximation in some applications (the *Born* approximation) is to assume that the diffracted wave is negligible compared to the incident wave. The Lippmann–Schwinger equation becomes

$$u_s(x) = k^2 \int_\Omega G(x,y)q(y)u_i(y)\, dy.$$

This time it is a *linear* integral equation for q, which is still an equation of the first kind; This inverse problem is therefore *ill posed*, as will be seen in example 2.12.

The nonlinear version of the problem is more difficult. It has been the subject of numerous works, both theoretical and numerical. Colton and Kress's work [COL 92] contains a very complete state of the art about this problem.

2.5. Other examples

EXAMPLE 2.12 (Gravimetric prospection).– The objective here is to determine the location, or the shape, of the anomalies of the gravity field in a known structure, from force measurements at the surface. Let Ω be a part of the Earth and ρ the density. The force due to gravity at a point $x \notin \Omega$ is given by Newton's law (G is the gravitational constant):

$$\phi(x) = \frac{G}{4\pi} \int_\Omega \frac{\rho(y)}{\|x-y\|^2}\, dy \qquad [2.24]$$

We shall confine ourselves this time to a 1D model, inspired by [KIR 96].

We want to determine the distribution $\rho(s)$, $0 \le s \le 1$, the mass density of an anomaly located at a depth h, from measurements of the vertical force $f(t)$.

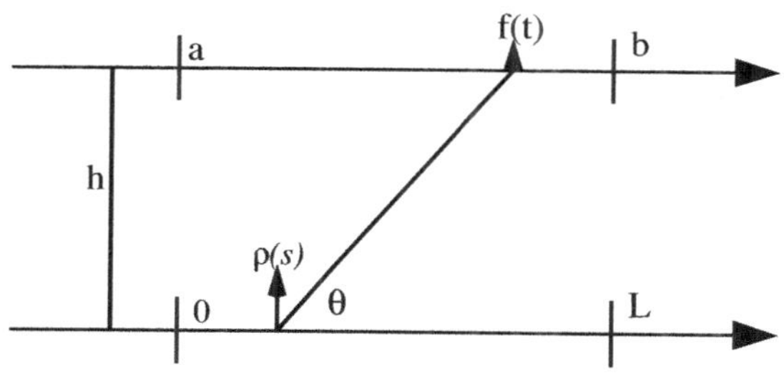

Figure 2.5. *Geometry of the prospecting gravimetric experiment*

The contribution to $f(t)$ due to the segment ds of the axis of the s is

$$G\frac{\sin\theta}{r^2}\rho(s)\, ds$$

where $r = \sqrt{h^2 + (s-t)^2}$. With $\sin\theta = h/r$, this yields

$$f(t) = G\int_0^L \frac{h}{(h^2+(t-s)^2)^{3/2}}\rho(s)\, ds \quad a \le t \le b \qquad [2.25]$$

The direct problem, which consists of calculating the force knowing the density distribution, is simply the evaluation of an integral. This inverse problem is the solution of an integral equation of the first kind. This is a problem similar to the differentiation mentioned in section 1.1, but with a general integral kernel, namely the function $(s,t) \mapsto \dfrac{h}{(h^2+(t-s)^2)^{3/2}}$. We will study this type of problem in more detail in Chapter 3 and this example is one of those that will be used as a guiding thread to illustrate the different concepts and tools coming across later in the book.

EXAMPLE 2.13 (Ray tracing).– This is a variant of the seismic model, in which a simplified propagation model is considered. We will follow the presentation of the book [GOR 91].

We will assume further that the geological model is stratified and in addition we will assume that the function $z \mapsto c(z)$ is increasing over the interval $[0, Z]$. We consider (this is Fermat's principle) that the waves travel as *rays* and that a fixed ray connecting a source (on the surface) to a receiver (also on the surface) is followed, as shown in Figure 2.6. This ray, after (possibly) turning back at depth $Z(p)$, emerges at point (Xp). The instant of emergence of this ray, noted $T(X)$, is measured.

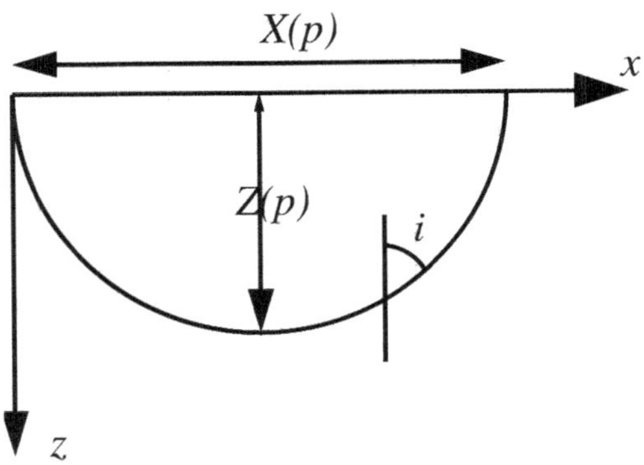

Figure 2.6. *Ray in a stratified soil*

It is shown (this is still Fermat's principle, see [AKI 80] and [GOR 91]) that the distance between the source and the receiver is given relatively to the *parameter of the ray* $p = \sin i/c(z)$ (which is a constant) by

$$X(p) = 2\int_0^{Z(p)} \frac{c(z)\,p}{\sqrt{1-c(z)^2\,p^2}}\,dz, \qquad [2.26]$$

similarly, the travel time is given by the integral

$$T(p) = 2\int_0^{Z(p)} \frac{1}{c(z)} \frac{1}{\sqrt{1 - c(z)^2 p^2}}\, dz, \qquad [2.27]$$

We have therefore obtained X and T based on p. There is no evidence that these functions are monotonic (moreover, there are counterexamples). The time delay function is then introduced and defined by:

$$\tau(p) = T(p) - pX(p) = 2\int_0^{Z(p)} \sqrt{c(z)^{-2} - p^2}\, dz,$$

which is monotonic. Then, $\tau'(p)$ is calculated in two different ways:

$$\tau'(p) = T'(p) - pX'(p) - X(p) = -2\int_0^{Z(p)} \frac{p}{\sqrt{c(z)^{-2} - p^2}}\, dz = -X(p)$$

so that we obtain $p = dT/dX$. This proves that p can be measured (it is the slope of the curve $X \mapsto T$).

The change in variables $\xi = c(z)^{-1}$ (the inverse function exists under the assumption that $z \mapsto c(z)$ is monotonic) gives:

$$\tau(p) = 2\int_{c_0^{-1}}^{p} \frac{\xi f(\xi)}{\sqrt{\xi^2 - p^2}}\, d\xi, \qquad [2.28]$$

where f is the inverse of the function $z \mapsto c^{-1}(z)$.

This equation is still an Abel integral equation, for which we have already seen that a simple inversion formula exists. We thus obtain

$$f(\xi) = -\frac{1}{\pi}\int_{c_0^{-2}}^{\xi} \frac{\tau'(p)}{\sqrt{\xi^2 - p^2}}\, dp \qquad [2.29]$$

Formula [2.29] shows that the solution of the inverse problem involves the calculation of the derivative of the function f. Once again, we meet the differentiation at the center of an inverse problem. If the data $T(X)$ are known only approximately, the calculation of the derivative of the inverse function of the derivative of $T(X)$ will be very inaccurate. The inverse problem is therefore likely to present instability.

EXAMPLE 2.14 (Inverse scattering).– This is a problem similar to that discussed in example 2.11, but this time the goal is to find the shape of the diffracting obstacle. This problem occurs, for example, in radar or sonar detection, which motivates a large number of studies. A sound wave with a given frequency is sent to an obstacle (represented by an open domain Ω), the reflected signal is measured and the objective is to recover the border of the obstacle. The equations are the same as in example 2.11, except that they are posed on the exterior of the obstacle, and that a boundary condition should therefore be added on its boundary (this can be a condition of the Dirichlet, Neumann or mixed type). We can write integral equations similar to [2.23] and it is shown that the diffracted wave has the asymptotic representation

$$u_s(x) = \frac{e^{ikx}}{\sqrt{|x|}}\left[u_\infty\left(\frac{x}{|x|}\right) + O\left(\frac{1}{x}\right)\right] \quad \text{when } |x| \to \infty \qquad [2.30]$$

The function u_∞ is called the *far field* and is what is measured. The inverse problem is therefore to determine the boundary of the obstacle knowing a measurement of the far field. This inverse problem is nonlinear. The reader is referred to [COL 92] for more information.

EXAMPLE 2.15 (Inverse spectral problems).– The motivation of these problems comes from a famous article by Marc Kac [KAC 66] *Can One Hear the Shape of a Drum?* The mathematical question posed by this article is to find the shape (the boundary) of a domain knowing its vibrational frequencies (that is the eigen frequencies of the Laplace operator on this domain). We know that the answer is in general negative, that is it is known how to build domains that are isospectral (they have the same spectrum), but not isometric. Examples of polygonal domains have been given by Driscoll [DRI 97]. An accessible introduction can be found in Kirch's work [KIR 96].

The examples presented in this chapter only scratch the surface of the domain. Many other examples can be found in works by Kirch [KIR 96], Engl *et al.* [ENG 96], Isakov [ISA 98] or by Ramm [RAM 05].

PART 2

Linear Inverse Problems

3

Integral Operators and Integral Equations

This chapter presents a brief introduction to integral operators, as well as to integral equations of the first kind. The latter constitute the most important example of linear inverse problems.

We will first present the main properties of integral operators in the context of L^2; in particular, the fact that if the kernel is square integrable, the associated operator is compact. We will then study the numerical approximation of equations of the first kind by the Galerkin method and the quadrature–collocation method.

We will not address the convergence of these approximations in this chapter: to this end, the natural context is the study of the regularizing properties of projection methods. For more details, the reader may refer to [KIR 96] and [KRE 89].

Finally, this study will be concluded in section 4.3 by the study of the spectral properties of the operators.

3.1. Definition and first properties

THEOREM 3.1.– Let K be a function in the space $L^2(]c,d[\times]a,b[)$. The formula

$$Au(t) = \int_a^b K(t,s)u(s)\,ds \quad t \in]a,b[, \qquad [3.1]$$

defines a linear continuous operator from $L^2(a,b)$ to $L^2(c,d)$.

PROOF.– The linearity is evident, only continuity (and the fact that Ku is an element of $L^2(c,d)$ if $u \in L^2(a,b)$, which will be an immediate consequence thereof) are to be proved.

Naturally, we want to bound

$$\int_c^d |Au(t)|^2 \, dt = \int_c^d \left(\int_a^b K(t,s)u(s) \, ds \right)^2 dt.$$

By the Cauchy–Schwarz inequality, we get:

$$\int_c^d |Au(t)|^2 \, dt \leq \int_c^d \left(\int_a^b |K(t,s)|^2 \, ds \right) \left(\int_a^b |u(s)|^2 \, ds \right) dt$$
$$\leq M^2 \int_a^b |u(s)|^2 \, ds, \quad [3.2]$$

(with $M^2 = \iint_{]c,d[\times]a,b[} |K(t,s)|^2 \, ds \, dt < \infty$, since $K \in L^2(]c,d[\times]a,b[)$), which proves that [3.1] defines correctly a continuous operator of $L^2(a,b)$ in $L^2(c,d)$ and at the same time shows that its norm is bounded by M. □

DEFINITION 3.1.– *The operator A defined in theorem 3.1 is called the integral operator with kernel K.*

EXAMPLE 3.1 (Volterra operators).– These are operators of the form

$$Au(t) = \int_0^t k(t,s)u(s) \, ds, \quad \text{for } t \in [0,1],$$

with $k \in L^2([0,1] \times [0,1])$.

In order to be able to use theorem 3.1, we must first work on a fixed interval. To this end, we use a well-known trick in integration theory: we introduce the characteristic function of the interval $[0,t]$, denoted by $\chi_{[0,t]}$. The definition of A becomes:

$$Au(t) = \int_0^1 \chi_{[0,t]}(s)k(t,s)u(s) \, ds.$$

To obtain a more symmetric definition, we observe that $0 \leq s \leq t \leq 1$ is equivalent to $(s,t) \in T$, where T is the lower triangle of the unit square $[0,1] \times [0,1]$. We thus obtain:

$$Au(t) = \int_0^1 \chi_T(s,t)k(s,t)u(s) \, ds.$$

We must therefore check that the function $(s,t) \to \chi_T(s,t)k(s,t) \in L^2([0,1] \times [0,1])$. However, it is clear that

$$\int_T |k(s,t)|^2 \, ds\, dt \leq \int_{[0,1]\times[0,1]} |k(s,t)|^2 \, ds\, dt < \infty.$$

This result can in particular be applied to example 1.1: it suffices to define $k(s,t) = 1, \forall (s,t)$.

EXAMPLE 3.2 (Gravimetric prospection).– example 2.12 is also in the realm of integral operators, as its kernel is defined by

$$K(s,t) = \frac{h}{(h^2 + (s-t)^2)^{3/2}}, \qquad [3.3]$$

and is regular, therefore square integrable, on any tile of the plane as soon as $h > 0$ (which is the case for a situation of physical interest).

A particularly simple class of integral operators is constituted of operators whose kernels are known as *degenerate*, that is of the form:

$$K(t,s) = \sum_{j=1}^{p} a_j(t) b_j(s).$$

The corresponding operators are of *finite rank*.

PROPOSITION 3.1.– Let A be an integral operator with degenerate kernel. The image of A is finite dimensional.

PROOF.– Let us prove that the image of A is generated by the functions $a_1, \ldots, a_p$. Its dimension is thus bounded by p.

Let $u \in L^2(a,b)$.

$$Au(t) = \int_a^b \sum_{j=1}^{p} a_j(t) b_j(s) u(s)\, ds = \sum_{j=1}^{p} \left(\int_a^b b_j(s) u(s)\, ds \right) a_j(t),$$

and this is indeed an element of the vector space $\text{span}\{a_1, \ldots, a_p\}$. □

This result will be used later to prove the compactness of the operator A.

PROPOSITION 3.2.– Let A be an integral operator with kernel K. The adjoint of A, A^*, is the integral operator with kernel K^*, with:

$$K^*(t,s) = K(s,t) \tag{3.4}$$

PROOF.– It suffices to start from the definition (see [A3.18]). Let $u \in L^2(a,b), v \in L^2(c,d)$.

$$(Au, v) = \int_c^d \left(\int_a^b K(t,s)u(s)\, ds \right) v(t)\, dt = \int_{]a,b[\times]c,d[} K(t,s)u(s)v(s)\, ds\, dt,$$

by Fubini's theorem. Exchanging again the integration order yields:

$$(Au, v) = \int_a^b \left(\int_c^d K(t,s)v(t)\, dt \right) u(s)\, ds = (u, A^*v),$$

according to the definition of the adjoint. By permuting the name of the variables, we obtain definition [3.4]. □

COROLLARY 3.1.– The integral operator A with kernel K is self-adjoint if, and only if, the kernel is symmetric:

$$K(s,t) = K(t,s), \quad \forall (s,t) \in [a,b] \times [c,d]. \tag{3.5}$$

The following result concerns the composition of operators.

PROPOSITION 3.3.– Let $A_1 \in \mathcal{L}(L^2(a,b), L^2(c,d))$ and $A_2 \in \mathcal{L}(L^2(c,d), L^2(e,f))$ be integral operators with kernels $K_1 \in L^2(]a,b[\times]c,d[)$ and $K_2 \in L^2(]c,d[\times]e,f[)$. The compound $A_2A_1 \in \mathcal{L}(L^2(a,b), L^2(e,f))$ is an integral operator with kernel

$$K(t,s) = \int_c^d K_2(t,r)K_1(r,s)\, dr. \tag{3.6}$$

PROOF.– Here again, we simply have to follow the definitions:

$$\begin{aligned}(A_2A_1)(u)(t) &= \textstyle\int_c^d K_2(t,r) \left(\int_a^b K_1(r,s)u(s)\, ds \right) dr \\ &= \textstyle\int_a^b \left(\int_c^d K_2(t,r)K_1(r,s)\, dr \right) u(s)\, ds,\end{aligned}$$

still by Fubini's theorem. It must still be verified that this kernel is indeed square integrable. First, by the Cauchy–Schwarz inequality,

$$\left(\int_c^d K_2(t,r)K_1(r,s)\, dr \right)^2 \le \int_c^d |K_2(t,r)|^2\, dr \int_c^d |K_1(r,s)|^2\, dr,$$

then by a double application of Fubini's theorem,

$$\int_{[a,]\times[e,f]} \left(\int_c^d K_2(t,r) K_1(r,s)\, dr \right)^2$$
$$\leq \int_{[a,]\times[c,d]} |K_1(r,s)|^2 \, dr\, ds \int_{[c,d]\times[e,f]} |K_2(t,r)|^2 \, dr\, dt.$$

□

It is usual to classify the integral equations that can be associated with the integral operator A into the following two categories:

– equations of the first kind: The equation is given as

$$Au = f, \quad \text{where } f \in L^2(c,d); \qquad [3.7]$$

– equations of the second kind: The equation is given as

$$u - Au = f, \quad \text{where } f \in L^2(c,d). \qquad [3.8]$$

This distinction is justified by the very different properties of these two types of equations for square-integrable kernels. Equations of the first kind, upon which we will concentrate our attention, result in ill-posed problems. In contrast, those of the second kind have, in general, a unique solution (this is due to the Fredholm alternative, stated in theorem A3.6). This distinction is related to the *compactness* of the operator. Indeed, we have the following result:

THEOREM 3.2.– Let $K \in L^2(]a,b[\times]c,d[)$. The integral operator A with kernel K is compact from $L^2(a,b)$ to $L^2(c,d)$.

PROOF.– We accept that it is possible to approximate the kernel K in the space $L^2(]a,b[\times]c,d[)$ by a sequence of degenerate kernels $(K_n)_{n\in\mathbf{N}}$ (see, for example, [BAU 87]). Denote by A_n the integral operator with kernel K_n. According to Proposition 3.1, A_n is a finite rank operator. Let us prove that the sequence A_n converges toward A. We obtain

$$\begin{aligned}\|(A_n - A)u\|_F^2 &= \int_c^d \left(\int_a^b (K_n(t,s) - K(t,s))u(s)\, ds\right)^2 dt \\ &\leq \left(\iint_{]a,b\times]c,d[} |K_n(t,s) - K(t,s)|^2 \, ds\, dt\right) \|u\|_E^2 \\ &= \|K_n - K\|_{L^2(]a,b[\times]c,d[)} \|u\|_E^2 .\end{aligned}$$

The first term tends to 0, according to the choice of K_n, which completes the proof.

□

By bringing together this result with corollary A3.3, which states that a compact operator cannot have a continuous inverse, we can see that integral equations of the first kind will always give rise to ill-posed problems. We will come back to this topic when we have introduced the singular value decomposition in Chapter 4.

Another way to understand the ill-posed nature of these equations is by means of the Riemann–Lebesgue lemma, which states that

$$\int_0^p iK(t,s)\sin(ns)\,ds \xrightarrow[n\to\infty]{} 0, \text{ in } L^2(0,1)$$

for any kernel $K \in L^2([0,1] \times [0,1])$ (this is a result on Fourier series). Thus, *high-frequency* perturbations (this is the case of $\sin(ns)$ for large n) are annihilated by the integration operation with a regular kernel. Such a perturbation is therefore "undetectable" from the perspective of solving the integral equation [3.7].

3.2. Discretization of integral equations

We will limit ourselves to briefly study two methods to discretize an integral equation: the quadrature–collocation method and the Galerkin method.

3.2.1. *Discretization by quadrature–collocation*

We start with a few reminders about quadrature formulas that consist of approximating an integral

$$I = \int_a^b \varphi(s)\,ds$$

by a weighted sum of the values of φ at points called *nodes*.

In the examples that follow, n is an integer. We will just present the simplest methods: the midpoint, trapezoid and Simpson rules.

EXAMPLE 3.3 (Midpoint rule).– The nodes of the midpoint rule are the points $s_{j-1/2} = s_j - h/2,\ j = 1\dots,n$, and the weights are all equal to h. We approximate the integral (the area under the curve) by the sum of the areas of the rectangles of width h and of height $\varphi(s_{j-1/2})$:

$$I^R = h\big(\varphi(s_{1/2}) + \cdots + \varphi(s_{j-1/2}) + \cdots + \varphi(s_{n-1/2})\big).$$

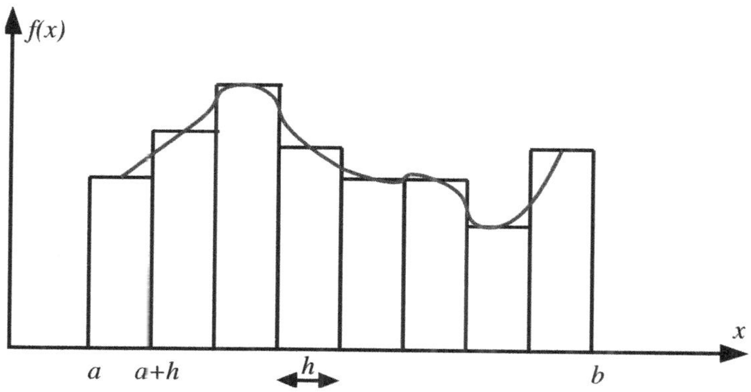

Figure 3.1. *Midpoint rule*

EXAMPLE 3.4 (Trapezoid rule).– The nodes of the trapezoid rule are the points s_j, $j = 1, \ldots, n$, where $h = (b - a)/n$. The weights are $w_j = h, j = 2, \ldots, n - 1$ as well as $w_1 = w_n = h/2$. The quadrature formula is obtained by replacing the integral by the sum of the areas of the trapezoids determined by the values of the function at the points s_j and s_{j+1}, which gives:

$$I^T = h\left(\frac{1}{2}\varphi(a) + \varphi(s_1) + \cdots + \varphi(s_{n-1}) + \frac{1}{2}\varphi(b)\right).$$

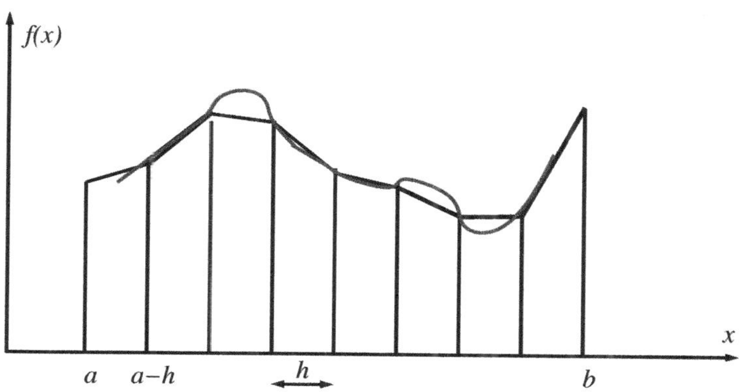

Figure 3.2. *Trapezoid rule*

EXAMPLE 3.5 (Simpson's rule).– This time it is assumed that n is even. The nodes of Simpson's rule are the same as for the trapezoid rule. The weights are

$w_{2j} = 2h/3$, $j = 1, \ldots, n/2 - 1$, $w_{2j+1} = 4h/3$, $j = 0, \ldots, n/2 - 1$, and $w_0 = w_n = h/3$. Simpson's quadrature formula is:

$$I^S = \frac{h}{3}\Big(\varphi(a) + 4\varphi(s_1) + 2\varphi(s_2) + \cdots + 2\varphi(s_{n-2}) + 4\varphi(s_{n-1}) + \varphi(b)\Big)$$

Under suitable hypotheses, the error obtained by replacing the integral by different quadrature formulas can be bounded.

PROPOSITION 3.4.– We have the following error bounds:

– Midpoint rule: Under the assumption $\varphi \in C^2[a,b]$,

$$\left|I - I^R\right| \leq \frac{(b-a)h^2}{24} \left\|\varphi''\right\|_\infty . \qquad [3.9]$$

– Trapezoid rule: Assuming $\varphi \in C^2[a,b]$,

$$\left|I - I^T\right| \leq \frac{(b-a)h^2}{12} \left\|\varphi''\right\|_\infty . \qquad [3.10]$$

– Simpson's rule: Assuming $\varphi \in C^4[a,b]$,

$$\left|I - I^S\right| \leq \frac{(b-a)h^4}{180} \left\|\varphi^{(4)}\right\|_\infty . \qquad [3.11]$$

PROOF.– Refer to a book on numerical analysis, for example [QUA 07] or [KRE 89]). □

In general, noting s_j and w_j for $j = 1, \ldots, n$, respectively, the nodes and the weights, a quadrature formula can be written as:

$$I^a = \sum_{j=1}^{n} w_j\, \varphi(s_j). \qquad [3.12]$$

The application of a quadrature formula to an integral equation (of the first kind) involves a collocation method. It is required that the integral equation is satisfied at a finite number of points $t_i, i = 1, \ldots, m$,

$$\int_a^b K(t_i, s)\, u(s)\, ds = f(t_i), \quad i = 1, \ldots, m, \qquad [3.13]$$

and the above integral is then replaced by the quadrature formula chosen in equation [3.12]. A (rectangular) system of equations is obtained:

$$\sum_{j=1}^{n} w_j K(t_i, s_j) u_j = f(t_i), \quad i = 1, \ldots, m. \qquad [3.14]$$

This is a linear system $A_h x = b$, as can be observed by defining

$$A_{hij} = w_j K(t_i, s_j),\ b_i = f(t_i,),\ x_j = u_j, \quad \text{for } j = 1, \ldots, n,\ i = 1, \ldots, m.$$

Where $m > n$, an overdetermined system is obtained. There is in any case no reason to choose $m = n$ and system [3.14] should generally be solved in the sense of least squares. We will return to these problems in Chapter 4 and a reminder on numerical methods is the subject of Appendix 1.

EXAMPLE 3.6.– In the case of the integral operator

$$Au(t) = \int_0^1 u(s)\, ds, \quad t \in [0, 1],$$

the matrix A_h is lower triangular, with $A_{ij} = h$, for $j \leq 1$, that is the system of equations [3.14] is:

$$h \sum_{j=1}^{i} u_j = f_i, \quad i = 1, \ldots, n.$$

It can be easily seen by induction that the solution is:

$$u_j = \frac{f_j - f_{j-1}}{h}, \quad j = 2, \cdots, n,$$

with $u_1 = f_1/h$, which is not surprising if we recognize a discretization of the first derivative. The inverse of A_h is effectively an approximation of the inverse of A.

3.2.2. *Discretization by the Galerkin method*

We now consider a projection method. The space $L^2(a, b)$ (respectively $L^2(c, d)$) is approximated by a sequence of subspaces of finite dimension E_n (respectively, F_m)

(it will be assumed $\dim E_n = n$, $\dim F_m = m$). Equation [3.7] is then projected on F_n, that is $u_n \in E_n$ is sought as a solution of the equation

$$(Au_n, v_n) = (f, v_m) \quad \forall v_m \in F_m. \qquad [3.15]$$

This equation is the Galerkin equation for u_n. In order to explain this equation, we introduce a basis $\{e_1, \ldots, e_n\}$ in the space E_n (respectively, a basis $\{f_1, \ldots, f_m\}$ in the space F_m). On this basis, u_n can be expanded in the form

$$u_n = \sum_{j=1}^{n} x_j e_j$$

and we choose $v_m = f_i$ in the Galerkin equation [3.15]. This yields:

$$\sum_{j=1}^{n} (Ae_j, f_i)\, x_j = (f, f_i), \quad i = 1, \ldots, m. \qquad [3.16]$$

This is still a (rectangular) system of equations, susceptible of numerical processing.

We should note a practical difference between the quadrature–collocation method in section 3.2.1 and the Galerkin method. In the quadrature–collocation method, the elements of the matrix and of the right-hand side are obtained simply as

$$A_{ij} = w_j\, K(t_i, s_j), \quad b_i = f(t_i) \qquad [3.17]$$

while the matrix elements of the Galerkin method are double integrals (simple for the vector):

$$A_{ij} = \iint_{]a,b[\times]c,d[} K(t,s)\, f_i(t)\, e_j(s)\, ds\, dt, \quad b_i = \int_c^d f(t) f_j(t)\, dt. \qquad [3.18]$$

The Galerkin method will result in a major additional cost compared to the quadrature–collocation method. On the other hand, it has improved convergence properties (higher order in other norms, see [KIR 96]).

EXAMPLE 3.7.– A simple, and useful example in practice, is provided by the choice of piecewise constant functions for the two approximation subspaces.

More specifically, we take $h_s = (b-a)/n,\ h_t = (d-c)/m$, and we subdivide the intervals $]a,b[$ and $]c,d[$ in n and m intervals of size h_s and h_t, respectively. We denote by $I_j^s =]a+(j-1)h_s, a+jh_s[$ and $I_i^t =]c+(i-1)h_t, c+ih_t[$ these intervals and define the basis functions by:

$$e_j(s) = \begin{cases} h_s^{-1/2}, & s \in I_j^s \\ 0 & \text{otherwise,} \end{cases} \quad j = 1, \ldots, n, \tag{3.19}$$

and

$$f_i(s) = \begin{cases} h_t^{-1/2}, & s \in I_i^t \\ 0 & \text{otherwise,} \end{cases} \quad i = 1, \ldots, m. \tag{3.20}$$

The elements of the matrix and of the right-hand side are then calculated by

$$A_{ij} = \frac{1}{\sqrt{h_s h_t}} \int_{I_i^t} \int_{I_j^s} K(t,s)\, dt\, ds,\ b_i = \int_{I_i^t} f(t)\, dt \tag{3.21}$$

We finish with a numerical example to underscore the difficulties of solving equations of the first kind.

EXAMPLE 3.8 (Gravimetric prospection).– We revisit our working example, the gravimetric prospection model, which we have met in example 2.12 in Chapter 2. We use equation [2.25], with $[a,b] = [0,1]$,$L = 2$, $h = 0.25$ and different values of n.

The exact solution is$f(t) = \sin(\pi t) + 1/2\sin(2\pi t)$, and the right-hand side is calculated accordingly. The equation is discretized by using the quadrature–collocation method discussed in section 3.2.1, with the midpoint rule. The resulting matrix is symmetric and positive definite.

We show in Figure 3.3 the results corresponding to the values of n equal to 20 and then to 40. Note the difference of scale on the vertical axis between the two figures. As can be seen, if the results are acceptable for $n = 20$, they are catastrophic for $n = 40$. This is obviously due to the ill-posed nature of the continuous problem. The more n increases, the more the discretization reproduces this ill-posed nature, which is numerically reflected by a huge condition number. The condition number of the matrix of this problem for different values of n is given in Table 3.1.

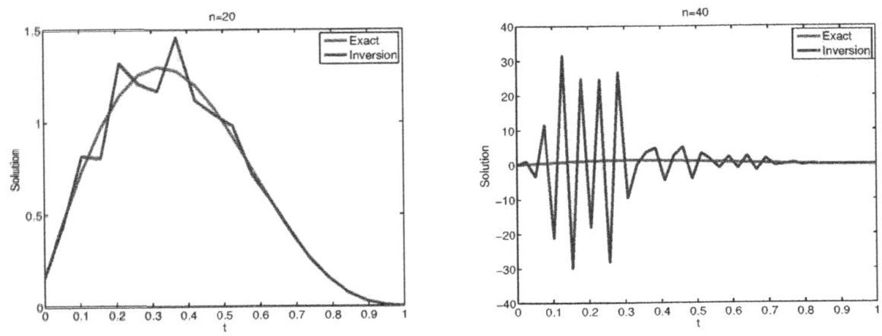

Figure 3.3. *Gravimetric prospection. Comparison of the exact solution with the computed solution. On the left $n = 20$, and on the right $n = 40$. For a color version of the figure, see www.iste.co.uk/kern/inverse.zip*

n	10	20	40	60	80	100
cond A	6.3×10^7	4.1×10^{16}	1.4×10^{19}	4.5×10^{19}	7.7×10^{19}	1.8×10^{20}

Table 3.1. *Geomagnetic prospection. Condition number of the matrix according to n*

3.3. Exercises

EXERCISE 3.1.– Integral operator example

Let A be the operator on $L^2(0,1)$ defined by

$$\forall u \in L^2(0,1), \ (Au)(t) = \int_0^t e^{-(t-s)} u(s)\, ds$$

1) Verify that A is effectively an integral operator and specify its kernel. How is the adjoint of A defined?

2) Show that for $f \in L^2(0,1)$, the following equivalence holds:

$$Av = f \Leftrightarrow \begin{cases} (e^t f(t))' = e^t v(t), \\ f(0) = 0 \end{cases}$$

Write a similar equivalence for equation $A^\star u = g$.

EXERCISE 3.2.– Backward heat equation

Consider the heat equation on a bounded interval (taken as $]0, \pi[$ in space to simplify, and as $]0, T[$ in time):

$$\begin{cases} \dfrac{\partial u}{\partial t} - \dfrac{\partial^2 u}{\partial x^2} = 0 & 0 < x < \pi,\ 0 < t < T \\ u(0,t) = u(\pi,t) = 0 & 0 < t < T \\ u(x,0) = u_0(x), & 0 < x < \pi \end{cases} \quad [3.22]$$

The inverse problem consists of finding the *initial condition* u_0 given a measurement of the solution at the final time $u_T(x) = u(x,T)$.

Let A be the (linear) operator defined by: $u_T = Au_0$. We recall (the proof is not required) that A is defined from $L^2(0,\pi)$ to $L^2(0,\pi)$.

It will be assumed during the whole exercise that the derivation under the $\sum$ symbol, or the inversion of the $\sum$ and $\int$ symbols are legitimate. Their justification is not required.

1) Solve equation [3.22], finding u in the form

$$u(x,t) = \sum_{n=1}^{\infty} a_n(t) u_n(x) \quad [3.23]$$

Let c_n by the Fourier coefficients of the initial condition u_0, given an ordinary differential equation satisfied by the function a_n. Specify the initial condition and solve this equation.

2) From the previous question, derive a representation of u_T as a series expansion:

$$u_T(x) = \sum_{n=1}^{\infty} a_n(T) u_n(x). \quad [3.24]$$

Also obtain an integral representation of the solution:

$$u_T(x) = \int_0^{\pi} K(x,y) u_0(y)\, dy.$$

Give the expression of the kernel K (in the form of a series expansion). Deduce that the operator A is self adjoint.

3) We denote by f_n the Fourier coefficients of the final condition u_T. Write the condition that these coefficients must satisfy in order for this function to be in the image of A.

4) It is assumed that the nth coefficient of u_T is known with a relative error δ. What is the relative error with respect to the norm of the solution u_0 (note $\tilde{u}_0$ the perturbed solution)?

Numerical application: Define $T = 1$ and it is assumed that $u_T = u_5$. Find the value of δ such that the relative error on u_0 be less than 100%.

4

Linear Least Squares Problems – Singular Value Decomposition

In this chapter, we study the main properties of linear inverse problems. We will work in the context of Hilbert spaces, so that the results apply (for example) to integral equations of the first kind, but we will mention the simplifications afforded by the finite dimensional case. We will then introduce the fundamental tool that the *singular value decomposition* (SVD) constitutes. This time, for the sake of simplicity of exposition, we will first consider the case of matrices, before presenting the case of operators, although the former is a special case of the latter. Finally, we will show how the SVD sheds light on the analysis of least squares problems.

This chapter is only concerned with mathematical aspects of least squares problems. We will deal specifically with methods for solving ill-posed problems in the following chapter.

In this chapter, we will denote by A a continuous linear operator from a Hilbert space E to a Hilbert space F: $A \in \mathcal{L}(E, F)$.

4.1. Mathematical properties of least squares problems

Given $\hat{z} \in F$, we search for $\hat{x} \in E$ a solution of:

$$A\hat{x} = \hat{z}. \qquad [4.1]$$

Let us return, in this particular case, to the discussion of Chapter 1 concerning well-posed and ill-posed problems:

– the operator A may not be surjective;

– it may not be injective;

– if an inverse exists, it may not be continuous.

As we have already said, the first difficulty is not serious: it suffices to limit ourselves to $\operatorname{Im} A$. The second is more troublesome: we must have the possibility to select, among several solutions, the solution that is appropriate to the problem. The last is fundamental for the applications and is linked to the fact that $\operatorname{Im} A$ may or may not be closed.

THEOREM 4.1.– Let $A \in \mathcal{L}(E, F)$, E and F two Hilbert spaces. Suppose that A be injective, and denote by $A^{-1} : \operatorname{Im} A \to E$ the inverse of A. There holds the following equivalence:

$$\operatorname{Im} A \text{ closed} \iff A^{-1} \text{ is continuous.}$$

PROOF.–

– for the direct implication, $W = \operatorname{Im} A$ is a Hilbert space (it is immediate that W is a pre-Hibertian space and it is complete because it is closed). The operator $\tilde{A} : E \to W$, $\tilde{A}u = Au$ for $u \in E$ is a continuous and bijective linear operator. A classical consequence of the open mapping theorem (theorem A3.2) is that $\tilde{A}^{-1}$ is continuous. The same holds for A^{-1};

– conversely, if A^{-1} is continuous, and $E = \operatorname{Im} A^{-1}$ is closed, $\operatorname{Im} A = (A^{-1})^{-1}(E)$ is closed in F. □

For the situations that we consider in this book, the operator may or may not be injective, but the usual situation will be that $\operatorname{Im} A$ is not closed (corollary A3.3 shows that if A is compact, then A has no continuous inverse, and in this case $\operatorname{Im} A$ will not be closed).

Problem [4.1] has a solution only if the right-hand side is in the image of A. As we have just seen, this condition may be too restrictive. We will return to this topic when we will have introduced the SVD, with Picard's condition (theorem 4.6). We are therefore looking for an alternative formulation of the original problem that allows for the notion of solution to be extended to a larger subspace.

We propose a formulation as a *least squares* problem: we thus replace [4.1] with:

$$\min_{x \in E} \frac{1}{2} \|Ax - \hat{z}\|_F^2 \qquad [4.2]$$

We will see that this problem is equivalent to a linear equation but for an operator different from A. We recall that A^* denotes *the adjoint* of A (the definition is recalled in Appendix 3, see equation [A3.18]). Equation [4.3] underneath is called *the normal equation* associated with [4.2]. We recall its important geometric interpretation after the proof of the theorem.

THEOREM 4.2.– Let $A \in \mathcal{L}(E, F)$, E, F two Hilbert spaces, and let $\hat{z} \in F$. An element $\hat{x} \in E$ is a solution of [4.2] if and only if

$$A^* A\hat{x} = A^* \hat{z} \qquad [4.3]$$

PROOF.– Equation [4.3] can be easily obtained by calculating the gradient of the functional $x \to \frac{1}{2}\|Ax - z\|^2$. We leave this method as an exercise for the reader. We will rather present the following elementary argument, borrowed from [BJÖ 96].

Let x satisfying [4.3]. For all $y \in E$:

$$\hat{z} - Ay = \hat{z} - Ax + A(x - y).$$

The normal equation [4.3] implies that the two terms of the sum are orthogonal (the residual $\hat{z} - Ax$ is orthogonal to the image of A). The Pythagorean theorem implies:

$$\|\hat{z} - Ay\|_F^2 = \|\hat{z} - Ax\|_F^2 + \|A(x - y)\|_F^2 \geq \|\hat{z} - Ax\|_F^2 .$$

x is therefore a solution of [4.2].

Conversely, let x such that $A^*(\hat{z} - Ax) = w \neq 0$. Choose $y = x + \varepsilon w$, with $\varepsilon > 0$. We then get:

$$\begin{aligned}\|\hat{z} - Ay\|_F^2 = (\hat{z} - Ay,\, \hat{z} - Ay) &= \|\hat{z} - Ax\|_F^2 - 2\varepsilon(\hat{z} - Ax,\, w) + \|Aw\|_F^2 \\ &< \|\hat{z} - Ax\|_F^2\end{aligned}$$

if ε is small enough. Therefore x is not a solution of [4.2]. □

REMARK 4.1.– The normal equation [4.3] can be rewritten as:

$$A^* \left(A\hat{x} - \hat{z}\right) = 0,$$

which expresses simply that the residual $\hat{z} - A\hat{x}$ is in the kernel of A^*, or that it is orthogonal to (the closure of) the image of A (see proposition A3.5). This leads to the well-known geometric illustration, recalled in Figure 4.1.

The solution of the least squares problem is such that Ax is the projection of $\hat{z}$ on the image of A.

Note that we have not yet mentioned the existence, nor the uniqueness, for the solutions of [4.2] (or of [4.3]). We have simply shown the equivalence of the two problems. The uniqueness is obviously linked to the injectivity of A, as stated in the following result.

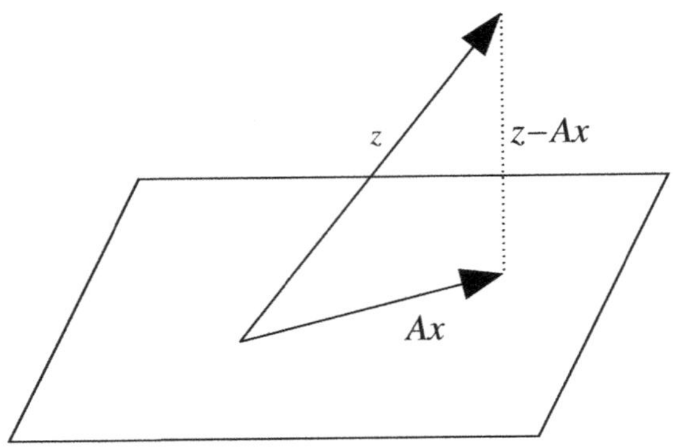

Figure 4.1. *Geometric illustration of the least squares*

LEMMA 4.1.– The solution of problem [4.2] is unique if and only if the operator A is injective.

PROOF.– Note first that $\operatorname{Ker} A^*A = \operatorname{Ker} A$. One direction is obvious. For the other, we have:

$$A^*Ax = 0 \Rightarrow (A^*Ax,\, x) = 0 \Rightarrow (Ax,\, Ax)_F = \|Ax\|_F = 0 \Rightarrow Ax = 0.$$

Consequently, A and A^*A are injective at the same time, which gives the result. □

With regard to the existence, we have the following result.

PROPOSITION 4.1.– Let $A \in \mathcal{L}(E, F)$, E and F two Hilbert spaces.

1) Equation [4.3] admits a solution if and only if $\hat{z} \in \operatorname{Im} A \oplus \operatorname{Im} A^{\perp}$;

2) If $\hat{z} \in \operatorname{Im} A \oplus \operatorname{Im} A^{\perp}$, the set S of solutions of [4.3] is a closed non-empty convex subset of E.

PROOF.–

1) Let $x \in E$ a solution of [4.3]. We thus get $Ax - \hat{z} \in \operatorname{Ker} A^* = \overline{\operatorname{Im} A}^{\perp} = (\operatorname{Im} A)^{\perp}$. Therefore, $\hat{z} = Ax + (\hat{z} - Ax) \in \operatorname{Im} A \oplus (\operatorname{Im} A)^{\perp}$.

Conversely, given $\hat{z} = z^1 + z^2$, with $z^1 \in \operatorname{Im} A$, $z^2 \in \operatorname{Im} A^{\perp}$. There thus exists $x \in E$, such that $Ax = z^1$. Obviously, $A^*Ax = A^*z^1$. But again because $(\operatorname{Im} A)^{\perp} = \operatorname{Ker} A^*$, we have $A^*z^2 = 0$, that is $A^*\hat{z} = A^*z^1 = A^*Ax$, and $\hat{z}$ is a solution of [4.3].

2) The solution set is non-empty according to point 1. It is an affine subspace, therefore it is in particular a convex set, and it is closed because it is the inverse image of $\{A^\star \hat{z}\}$ by the continuous operator $A^\star A$. This shows that $S = x_0 + \operatorname{Ker} A$, where x_0 is any solution of [4.3]. □

LEMMA 4.2.– The subspace $\operatorname{Im} A + \operatorname{Im} A^\perp$ is dense in F.

PROOF.– If $x \in (\operatorname{Im} A + \operatorname{Im} A^\perp)^\perp$, then, for all $y \in \operatorname{Im} A$ and $z \in \operatorname{Im} A^\perp$, we have $(x, y + z) = 0$. By choosing first $z = 0$, we obtain $x \in \operatorname{Im} A^\perp$, then by choosing $y = 0$, we obtain $x \in \operatorname{Im} A^{\perp\perp} = \overline{\operatorname{Im} A}$. In other words, $(\operatorname{Im} A + \operatorname{Im} A^\perp)^\perp = \overline{\operatorname{Im} A} \cap \operatorname{Im} A^\perp = \{0\}$, which is equivalent to the desired density. □

We thus have been successful in extending the notion of solution to a dense subspace in F, which is "almost as good as" extending it to the whole of F.

REMARK 4.2.– It should be noted that, in the general case, the condition $\hat{z} \in \operatorname{Im} A \oplus \operatorname{Im} A^\perp$ is *non-trivial*. Indeed, the projection theorem only states $F = \overline{\operatorname{Im} A} \oplus \operatorname{Im} A^\perp$, which is different if $\operatorname{Im} A$ is not closed.

Here, we find again the significance of this condition. In this cases, and only in this case, problem [4.3], and therefore [4.2], *always* admits a solution. In the finite dimensional case, this condition is obviously automatically satisfied (any subspace is closed). We will meet this result again in proposition 4.2.

COROLLARY 4.1.– If $\hat{z} \in \operatorname{Im} A \oplus \operatorname{Im} A^\perp$, problem [4.2] admits a unique minimal norm solution.

PROOF.– Denote S the set of solutions of [4.2]. Looking for a minimal norm solution for this least squares problem is equivalent to solve the following problem:

$$\min_{x \in S} \|x\|_E, \quad S = \{x \in E, \ \|Ax - b\|_F \text{ minimal}\}$$

that is to project the origin on the set S. According to proposition 4.1, S is a non-empty closed convex subset of E. The projection theorem A3.1 implies that S has a minimal norm element, which is the solution searched for. □

We will note $\hat{x}$ this particular solution.

REMARK 4.3.– It is easy to see that $\hat{x}$ depends linearly on $\hat{z}$ (e.g. because of the normal equation). The linear operator $A^\dagger : \operatorname{Im} A \oplus \operatorname{Im} A^\perp \mapsto E$ defined by $A^\dagger \hat{z} = \hat{x}$ is called the *pseudo-inverse* of A (see, for example, [GOL 96] for its algebraic properties). It can be shown (see, for example, [ENG 96]) that $A^\dagger$ is continuous if and only if $\operatorname{Im} A$ is closed.

EXAMPLE 4.1 (Canonical injection).– Note A the canonical injection from $H_0^1(0,1)$ to $L^2(0,1)$ ($Au = u$, for $u \in H_0^1(0,1)$). Inverting A amounts in fact to differentiate a function and we can therefore expect some difficulties. We can already note that the image of A is (obviously) $H_0^1(0.1)$, viewed as a subspace of $L^2(0,1)$, which is not closed (it is dense, and strictly included in $L^2(0,1)$). Obviously, in this case, A is injective.

We start by determining the adjoint of A, and to this end we must specify the norms. We choose as norm on $H_0^1(0,1)$ the quantity $\|u\|_1^2 = \int_0^1 |u'(x)|^2\, dx$ (this is indeed a norm because of the Poincaré inequality). Under these conditions, for $v \in L^2(0,1)$, the adjoint A^*v is characterized by:

$$(u,\, A^*v) = (Au,\, v), \quad \forall u \in H_0^1(0,1),$$

that is, by writing out the inner products:

$$\int_0^1 u'(x)(A^*v)'(x)\, dx = \int_0^1 u(x)v(x)\, dx, \quad \forall u \in H_0^1(0,1).$$

In this last equality, we can recognize the variational formulation of a second-order elliptic problem. A^*v is the unique solution (according to the Lax–Milgram lemma) of the Dirichlet problem:

$$\begin{cases} -(A^*v)'' = v, & x \in]0,1[\\ (A^*v)(0) = (A^*v)(1) = 0. \end{cases} \qquad [4.4]$$

Thus, the adjoint of A is calculated by integrating a differential equation. In particular, the image of A^* consists of functions of $H_0^1(0,1) \cap H^2(0,1)$, therefore of more regular functions.

Given $\hat{z} \in L^2(0,1)$, the least squares problem [4.2] has a solution only if $z \in H_0^1(0,1)$, and the solution is then $u = \hat{z}$. If this condition is not satisfied, the minimization problem has no solution (the minimum is not reached).

4.1.1. *Finite dimensional case*

If E and F are two finite-dimensional spaces, it can be assumed that $E = \mathbf{R}^n$ and $F = \mathbf{R}^m$. In this case, A can be identified to a matrix $A \in \mathbf{R}^{m\times n}$. We will assume in general that $m > n$, that is the problem is *overdetermined*.

In this case, the results of the previous section can be made precise.

PROPOSITION 4.2.– When E and F are two finite-dimensional spaces, the least squares problem always admits at least one solution.

This solution is unique if and only if A is of maximal rank (of rank n if the problem is overdetermined).

PROOF.– This is a simple application of lemma 4.1 and proposition 4.1, since the image of A is always closed. A is injective if and only if it is of maximal rank (when $m > n$).

A direct proof can also be given, not using the results of the previous section. We show that the normal equations always have at least a solution, that is to say that $A^T\hat{z} \in \mathrm{Im}(A^T)$. To this end, we shall begin by observing that $\mathrm{Im}\, A^T = \mathrm{Im}(A^T A)$. In effect, $\mathrm{Im}(A^T A) = \mathrm{Ker}(A^T A)^\perp = \mathrm{Ker}(A)^\perp = \mathrm{Im}(A^T)$. It thus suffices to verify that $A^T\hat{z} \in \mathrm{Ker}(A)^\perp$; however, this is immediate, since, if $x \in \mathrm{Ker}(A)^\perp$, then $(A^T\hat{z}, x) = (\hat{z}, Ax) = 0$. □

PROPOSITION 4.3.– Under the assumption that A is of rank n, the matrix of normal equations $A^T A$ is *positive definite*.

PROOF.– It is easy to see that $A^T A$ is positive semidefinite (this is in fact always true):

$$(A^T Ax, x) = (Ax, Ax) = \|x\|^2 \geq 0.$$

In addition, when A is of rank n, we have seen (in lemma 4.1) that the (square) matrix $A^T A$ is injective, thus invertible. It is therefore definite positive. □

This result might suggest that a method to solve [4.2] is to invert [4.3] by the Cholesky method. We recall in Appendix 1 why this is not necessarily a good idea.

EXAMPLE 4.2 (Linear regression).– We aim to draw a straight line $y(t) = \alpha + \beta t$ through a set of experimental points $(t_i, y_i), i = 1, \ldots, m$. This leads to the overdetermined system

$$\begin{pmatrix} 1 & t_1 \\ 1 & t_2 \\ \vdots & \vdots \\ 1 & t_m \end{pmatrix} \begin{pmatrix} \alpha \\ \beta \end{pmatrix} = \begin{pmatrix} 1 & y_1 \\ 1 & y_2 \\ \vdots & \vdots \\ 1 & y_m \end{pmatrix}.$$

The normal equation

$$\begin{pmatrix} m & \sum_{i=1}^m t_i \\ \sum_{i=1}^m t_i & \sum_{i=1}^m t_i^2 \end{pmatrix} \begin{pmatrix} \alpha \\ \beta \end{pmatrix} = \begin{pmatrix} \sum_{i=1}^m y_i \\ \sum_{i=1}^m y_i t_i \end{pmatrix}$$

has the solution

$$\beta = \frac{\displaystyle\sum_{i=1}^{m} y_i t_i - m\overline{y}\overline{t}}{\displaystyle\sum_{i=1}^{m} t_i^2 - m\overline{t}^2}, \qquad \alpha = \overline{y} - \beta\overline{t},$$

where we have denoted by $\overline{y} = \sum_{i=1}^{m} y_i/m$ and $\overline{t} = \sum_{i=1}^{m} t_i/m$.

Note that the line obtained passes through the means $(\overline{t}, \overline{y})$.

4.2. Singular value decomposition for matrices

The singular value decomposition (SVD) has become for a few decades a fundamental tool to investigate a growing number of linear problems. The decomposition was discovered more than a hundred years ago by Beltrami, but only became a numerical tool since the end of the 1960s, when Golub showed how it could be computed in a stable and (reasonably) effective manner.

This decomposition is a kind of diagonalization that gives a complete "identity card" of the operator. We will see in particular that it gives a simple (in theory) solution to the least squares problem.

We state the main theorem in the case of real matrices.

THEOREM 4.3.– Let $A \in \mathbf{R}^{m\times n}$ be a matrix of rank r. There exists two orthogonal matrices $U \in \mathbf{R}^{m\times m}$ ($U^TU = UU^T = I_m$) and $V \in \mathbf{R}^{n\times n}$ ($V^TV = VV^T = I_n$) such that

$$A = U\Sigma V^T, \qquad \Sigma = \begin{pmatrix} \Sigma_1 & 0 \\ 0 & 0 \end{pmatrix}, \tag{4.5}$$

where $\Sigma \in \mathbf{R}^{m\times n}$, $\Sigma_1 = \operatorname{diag}(\sigma_1, \sigma_2, \dots, \sigma_r)$, and

$$\sigma_1 \geq \sigma_2 \geq \dots \geq \sigma_r > 0.$$

Component by component, the matrix identity [4.5] becomes:

$$Av_j = \sigma_j u_j, \quad A^T u_j = \sigma_j v_j, \qquad \text{for } j = 1, \dots, n, \tag{4.6}$$

$$A^T u_j = 0, \qquad \text{for } j = n+1, \dots, m. \tag{4.7}$$

If we denote by $U = (u_1, \cdots, u_m)$, $V = (v_1, \cdots, v_n)$ the columns of the matrices U and V, the vectors u_j and v_j are, respectively, the right and left *singular vectors* associated with the *singular value* σ_j.

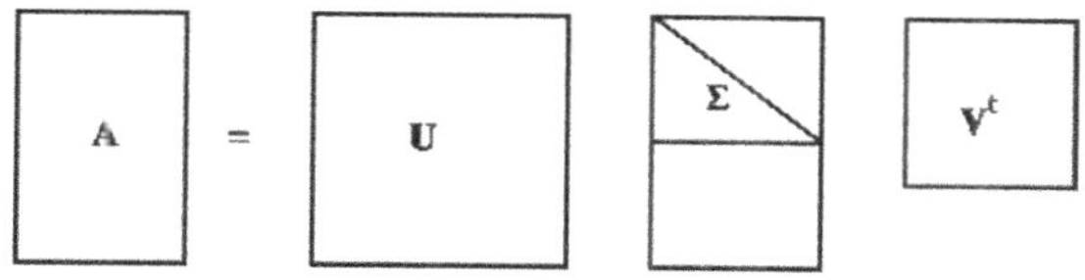

Figure 4.2. *Illustration of the SVD*

PROOF.– The proof is carried out by induction on n.

By the definition of a subordinate matrix norm, there exists a vector $v_1 \in \mathbf{R}^n$ such that

$$\|v_1\|_2 = 1, \quad \|Av_1\|_2 = \|A\|_2 \underset{\text{def}}{=} \sigma,$$

where σ is strictly positive (if $\sigma = 0$, then $A = 0$, and there is nothing to prove). Let $u_1 = 1/\sigma\, Av_1 \in \mathbf{R}^m$. We complete vector v_1 to an orthonormal basis of $\mathbf{R}^n$, and we note $V = (v_1, V_1) \in \mathbf{R}^{n\times n}$ the matrix formed by the basis vectors. We do the same for u_1 and $\mathbf{R}^m$, noting $U = (u_1, U_1) \in \mathbf{R}^{m\times m}$. It should be observed that matrices U and V are orthogonal by construction.

According to our choice of U_1, $U_1^T A v_1 = \sigma U_1^T u_1 = 0$, and so the product $U^T AV$ has the following block structure:

$$A_1 \underset{\text{def}}{=} U^T AV = \begin{pmatrix} \sigma & w^T \\ 0 & B \end{pmatrix}$$

with $w^T = u_1^T AV_1$ and $B = U_1^T AV_1 \in \mathbf{R}^{(m-1)\times(n-1)}$.

Since U and V are orthogonal, $\|A_1\|_2 = \|A\|_2 = \sigma$. But the double inequality

$$\|A_1\|_2 \left(\sigma^2 + w^T w\right)^{1/2} \geq \left\| A_1 \begin{pmatrix} \sigma \\ w \end{pmatrix} \right\|_2 = \left\| \begin{pmatrix} \sigma^2 + w^T w \\ Bw \end{pmatrix} \right\|_2 \geq \sigma^2 + w^T w,$$

shows that $\|A_1\|_2 \geq (\sigma^2 + w^T w)^{1/2}$. Therefore, we must have $w = 0$. We can then finish the proof by applying the induction hypothesis to B. □

From the geometric point of view, this theorem expresses that any linear mapping can be seen, after a change in *orthonormal* bases in each of the spaces, as acting by dilation in each direction. We insist on the significance of the fact that the considered bases are orthonormal. This constitutes the main difference with diagonalization. Indeed, if any matrix has a SVD, only normal matrices (these are the matrices that commute with their adjoint) are diagonalizable in an orthonormal basis.

EXAMPLE 4.3.– We illustrate this geometric aspect in Figure 4.3, with the example of matrix

$$A = \begin{pmatrix} 3 & 1 \\ 1 & 3 \end{pmatrix}$$

whose SVD is (exercise):

$$A = \begin{pmatrix} 1/\sqrt{2} & -1/\sqrt{2} \\ 1/\sqrt{2} & 1/\sqrt{2} \end{pmatrix} \begin{pmatrix} 4 & 0 \\ 0 & 2 \end{pmatrix} \begin{pmatrix} 1/\sqrt{2} & -1/\sqrt{2} \\ 1/\sqrt{2} & 1/\sqrt{2} \end{pmatrix}^T$$

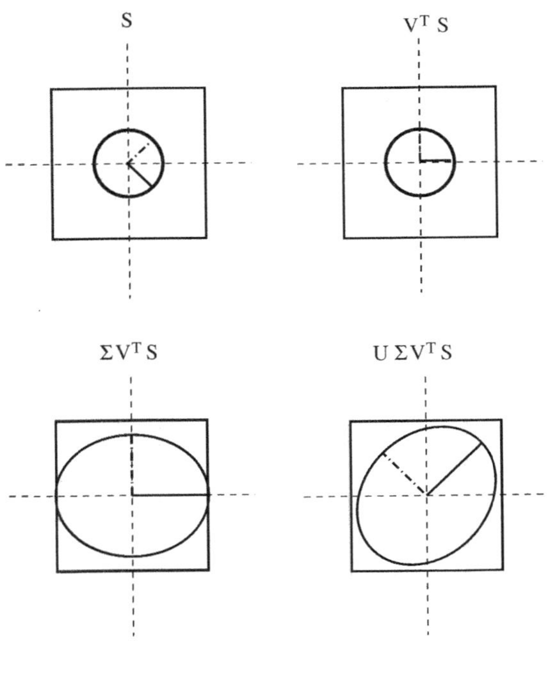

Figure 4.3. *SVD geometric significance*

We calculate the image of the unit circle. We start with the action of V^T, which is a rotation (of angle $-45°$), then by Σ, whose action expands the x-axis by a ratio of 4 and the y-axis by a ratio of 2, finally by U, which is a rotation of angle $45°$.

The singular values are related to the eigenvalues of matrices "derived" from A. The following proposition follows easily from the SVD of A.

PROPOSITION 4.4.– Let $A \in \mathbf{R}^{m\times n}$ a matrix. Denote by $A = U\Sigma V^T$ its SVD.

1) The eigenvalues of the matrix A^TA are the numbers σ_j^2, $j = 1, \dots, n$, and its eigenvectors are the left singular vectors of A, v_j, $j = 1, \dots, n$;

2) The eigenvalues of the matrix $\begin{bmatrix} 0 & A^T \\ A & 0 \end{bmatrix}$ are the numbers $\pm\sigma_j$, $j = 1, \dots, n$, and its eigenvectors are $\dfrac{1}{\sqrt{2}}\begin{bmatrix} v_j \\ \pm u_j \end{bmatrix}$

This proposition allows us to clarify in what sense the SVD of a matrix A is unique. Since the singular values are the eigenvalues of A^TA, they are determined by A, and therefore unique. The first n right singular vectors belonging to a simple singular value are also unique (up to a factor ± 1); on the other hand, for a multiple singular value, only the subspace is unique. Finally, the left singular vectors corresponding to a simple singular value are also unique. On the other hand, the last $m - n$ right singular vectors are not determined only by A, only the subspace generated by $u_{n+1}, \dots, u_m$ is unique.

It may be convenient to write the SVD of A in a form slightly different from [4.5]:

– first, by defining $U_n = (u_1, \dots, u_n)$, we get

$$A = U_n \Sigma V^T \qquad [4.8]$$

sometimes referred to as the "singular value factorization" of A;

– similarly, if A is not of rank n, we denote by $U_r = (u_1, \dots, u_r)$, $V_r = (v_1, \dots, v_r)$ the matrices formed by the first r singular vectors and $\Sigma_r = \operatorname{diag}(\sigma_1, \dots, \sigma_r)$. Then:

$$A = U_r \Sigma_r V_r^T = \sum_{i=1}^{r} \sigma_i u_i^T v_i. \qquad [4.9]$$

Note that matrices U_r and V_r are orthogonal, since this is the case of the vectors $u_1, \dots, u_r$ and $v_1, \dots, v_r$. The decomposition [4.9] corresponds to a sum of matrices of rank 1.

This latter form is the one that best lends itself to the generalization to operators in Hilbert spaces that we will cover in section 4.3. It expresses that the columns of V_r

form a basis of $(\mathrm{Ker}\, A)^{\perp}$, that those of U_r form a basis of $\mathrm{Im}\, A$, and that by expanding the vectors in this basis, we get

$$x = x^0 + \sum_{i=1}^{r} (x,\, v_i)\, v_i \Longrightarrow Ax = \sum_{i=1}^{r} \sigma_i (x,\, v_i)\, u_i, \quad \text{where } x^0 \in \mathrm{Ker}\, A. \qquad [4.10]$$

As we have mentioned, the main properties of a matrix can be read from its SVD.

PROPOSITION 4.5.– We have the following relationships:

1) The rank of A is the equal to the number of non-zero singular values;

2) $\mathrm{Ker}\, A = \mathrm{span}(v_{r+1}, \cdots, v_n)$, $\quad \mathrm{Im}\, A = \mathrm{span}(u_1, \cdots, u_r)$;

3) $\mathrm{Ker}\, A^T = \mathrm{span}(v_1, \cdots, v_r)$, $\quad \mathrm{Im}\, A^T = \mathrm{span}(u_{r+1}, \cdots, u_m)$;

4) $\|A\|_2 = \sigma_1$.

PROOF.– Matrices U and V are isometries and thus bijections. The different properties are therefore consequences that they are true for Σ:

1) It is clear that the rank of a diagonal matrix is equal to the number of diagonal non-zero elements. The general result is a consequence of the previous observation.

2) Given $v \in \mathrm{Ker}\, A$, then $V^T v \in \mathrm{Ker}\, U^T A V = \mathrm{Ker}\, \Sigma$, and vice versa. Again, it is clear that the kernel of Σ is spanned by columns $r+1$ to n of the identity, and thus, the kernel of A is created by the product of V by these columns, that is the vector $v_{r+1}, \ldots, v_n$.

A similar reasoning gives the result for the image of A.

3) The reasoning is identical to that of the previous point.

4) Since U and V are isometries, we get for $x \in R^n$:

$$\frac{\|Ax\|_2}{\|x\|_2} = \frac{\|U\Sigma V^T x\|_2}{\|x\|_2} = \frac{\|\Sigma y\|_2}{\|y\|_2}$$

with $y = V^T x$ (and thus $\|y\|_2 = \|x\|_2$). Therefore:

$$\frac{\|Ax\|_2}{\|x\|_2} = \frac{(\sum_{j=1}^{r} \sigma_j^2 y_j^2)^{1/2}}{(\sum_{j=1}^{r} y_j^2)^{1/2}} \leq \sigma_1,$$

with equality for $y = (1, 0, \ldots, 0)^T$. □

The SVD allows us to establish a result about the approximation of a matrix by matrices of lower rank, which will be useful when we will study spectral truncation (section 5.2.2).

PROPOSITION 4.6.– For $k = 1, \dots, n-1$, denote $A_k = U\Sigma_k V^T$.

Then, $\|A - A_k\|_2 = \sigma_{k+1}$, and matrix A_k is the matrix of rank k that minimizes the difference $\|A - B\|_2$.

PROOF.– It is clear that A_k is of rank k.

Since $\Sigma - \Sigma_k = \operatorname{diag}(0, \dots, 0, \sigma_{k+1}, \dots, \sigma_n)$, we have

$$\|A - A_k\|_2 = \left\|U(\Sigma - \Sigma_k)V^T\right\|_2 = \|\Sigma - \Sigma_k\|_2 = \sigma_{k+1}.$$

Finally, let B be a matrix of rank k. The kernel of B is thus of dimension $n-k$ and the space spanned by $v_1, \dots, v_{k+1}$ is of dimension $k+1$. The sum of the dimensions is equal to $(n-k)+(k+1) > n$ and as a result these two subspaces have a non-trivial intersection. Let x be a vector of norm 1 in the intersection. On the one hand:

$$\|A - B\|_2 \geq \|(A - B)x\|_2$$

and on the other hand ($(A - B)x = Ax$, since $x \in \operatorname{Ker} B$):

$$\|(A - B)x\|_2 = \|Ax\|_2 = \left\|U\Sigma V^T x\right\|_2 = \left\|\Sigma(V^T x)\right\|_2 \geq \sigma_{k+1}\left\|V^T x\right\|_2 = \sigma_{k+1},$$

hence the result by comparing the two inequalities. □

4.3. Singular value expansion for compact operators

In this section, we will assume that A is a compact operator from a Hilbert space E to a Hilbert space F (see section A3.2.3). We are going to generalize the SVD to this situation. The main difference will be the existence of a (countable) *infinity* of singular values . As a result, we will obtain a criterion for the existence of a solution to the least squares problem (Picard's criterion, theorem 4.6).

THEOREM 4.4.– Let $A : E \to F$ a *compact* as the operator. There exists a *decreasing* sequence $(\sigma_j)_{j\in\mathbf{N}} \in \mathbf{R}_+$, and two orthonormal families $(e_j)_{j\in\mathbf{N}} \in E$, $(f_j)_{j\in\mathbf{N}} \in F$ such that:

1) $Ae_j = \sigma_j f_j$; $A^* f_j = \sigma_j e_j$, $j \in \mathbf{N}$;

2) For all $x \in E$, we have the expansion:

$$x = x_0 + \sum_{j=1}^{+\infty} (x, e_j)e_j, \quad \text{where } x_0 \in \operatorname{Ker} A; \qquad [4.11]$$

3) For all $x \in E$ and $y \in F$, we have:

$$Ax = \sum_{j=1}^{+\infty} \sigma_j (x, e_j) f_j, \qquad A^* y = \sum_{j=1}^{+\infty} \sigma_j (y, f_j) e_j \tag{4.12}$$

The sequence $(e_j)_{j \in \mathbf{N}}$ is a Hilbert basis of of $\operatorname{Ker} A^{\perp}$, and the sequence $(f_j)_{j \in \mathbf{N}}$ is a Hilbert $\overline{\operatorname{Im} A}$.

PROOF.– Consider the self-adjoint operator $T = A^* A$. T is compact the composition of A compact and of A^* continuous (and compact). The spectral theory of compact self-adjoint operators (see theorem A3.7 or [BRE 11]) implies that T has a sequence of non-zero eigenvalues $(\lambda_j)_{j \in \mathbf{N}} \in \mathbf{R}^*$ and eigenvectors $(e_j)_{j \in \mathbf{N}}$, such that

$$Te_j = \lambda_j e_j$$

and if $x \in E$, there exists $x_0 \in \operatorname{Ker} T$, such that $x = x_0 + \sum_{j=1}^{+\infty} (x, e_j) e_j$.

Assume that the $(\lambda_j)_{j \in \mathbf{N}}$ are ordered in a decreasing sequence, each eigenvalue being counted according to its multiplicity. Then, observe that λ_j is strictly positive, for any j, since:

$$\lambda_j = (Te_j, e_j) = (A^* A e_j, e_j) = \|Ae_j\|^2 \geq 0$$

and λ_j is non-zero by hypothesis. Therefore, we can define $\sigma_j = \lambda_j^{1/2}$, $\sigma_j > 0$ and let $f_j = \dfrac{1}{\sigma_j} A e_j \in \operatorname{Im} A \subset F$.

Then, we see that $A^* f_j = \dfrac{1}{\sigma_j} A^* A e_j = \frac{1}{\sigma_j} T e_j = \sigma_j e_j$ and $(f_j)_{j \in \mathbf{N}}$ is an orthonormal sequence in F, since:

$$(f_j, f_k) = \frac{1}{\sigma_j \sigma_k} (Ae_j, Ae_k) = \frac{\lambda_k}{\sigma_j \sigma_k} (e_j, e_k) = \sqrt{\frac{\sigma_k}{\sigma_j}} \delta_{jk} = \delta_{jk}.$$

The identities in (1) are immediate. In order to verify the representation [4.11], it should first be observed that $\operatorname{Ker} T = \operatorname{Ker}(A^* A) = \operatorname{Ker} A$. One inclusion is trivial. For the other:

$$A^* A x = 0 \Rightarrow (A^* A x, x) = \|Ax\|^2 = 0 \Rightarrow x \in \operatorname{Ker} A.$$

The reasoning is the same as in proposition 4.3. The expansion [4.11] is then that is obtained above.

To conclude, we will prove the first expansion in [4.12]. For $x \in E$, let

$$X = \sum_{j=1}^{+\infty} \sigma_j (x, e_j) f_j.$$

The series converges in E, with $\|X\| \leq \sigma_1 \|x\|$, and $X \in \overline{\operatorname{Im} A}$. As a result, $X - Ax \in \operatorname{Ker} A^* \cap \overline{\operatorname{Im} A} = \operatorname{Ker} A^* \cap (\operatorname{Ker} A^*)^\perp$ and thus $X = Ax$. The second expansion is established in the same way. □

DEFINITION 4.1.– *The quantities σ_j are called the singular values of A. Vectors e_j and f_j are the singular vectors. The expansion obtained in* [4.11] *is called the singular value expansion, also denoted by SVE, of A.*

REMARK 4.4.– We have assumed, for simplicity, that the operator A^*A has an infinite number of eigenvalues. The case of a finite number of eigenvalues exists but does not correspond to situations found in practice. We have chosen to eliminate this case to obtain a simpler statement.

EXAMPLE 4.4 (Canonical injection).– We shall determine for the singular value expansion of the operator of example 4.1. According to [4.4], the singular values and the associated functions are a solution of

$$\begin{cases} e_j = \sigma_j f_j, \\ -\sigma e_j'' = f_j, \quad \text{in }]0,1[, e_j(0) = e_j(1) = 0. \end{cases} \qquad [4.13]$$

By solving the system, we can see that the singular values are the quantities $\sigma_k = 1/(k\pi)$, $k \geq 1$, and the eigenfunctions are given by:

$$e_j(t) = \sqrt{2}\sin(k\pi t), \quad f(j) = \sqrt{2}k\pi \sin(k\pi t).$$

The interpretation of the expansions [4.11] and [4.12] is given as

$$u = \sum u_j \sqrt{2}\sin(k\pi t), \text{ with } u_j = \sqrt{2}\int_0^1 u(t)\sin(k\pi t)\, dt.$$

We recover the expansion of u in Fourier series (in the sense of $L^2(0,1)$).

EXAMPLE 4.5 (Gravimetric prospection).– We return to our guiding example, initiated in example 2.12. This example is not susceptible of an exact calculation, we will follow a numerical approach. In this example, and in the following, we will use the MATLAB *Regularization Toolbox*, written by Hansen [HAN 07], and is freely available[1]. We will use it to illustrate the concepts seen in this chapter and in the following chapter. This example corresponds to the `geomag` function of the Toolbox.

We use the default values for the parameters, that is $[a, b] = [0, 1]$, $L = 1, h = 0.25$, and the exact solution $f(t) = \sin(\pi t) + 0.5\sin(2\pi t)$. The operator is discretized by the quadrature collocation method (see section 3.2.1), leading here to a square matrix A (because the range interval is the same as the domain interval), of size n. Figure 4.4 represents the singular values of the operator. An exponential decrease in the singular values can be observed, which is linked to the very smooth nature of the kernel. This decay stops at the level of the "machine precision", approximately 10^{-16}. Figure 4.5 represents the first nine singular vectors. It can be observed that they have the same oscillating nature as the eigenfunctions in the previous example. The vectors corresponding to the singular values of highest rank (hence the smallest singular values, see Figure 4.4) are increasingly more oscillatory.

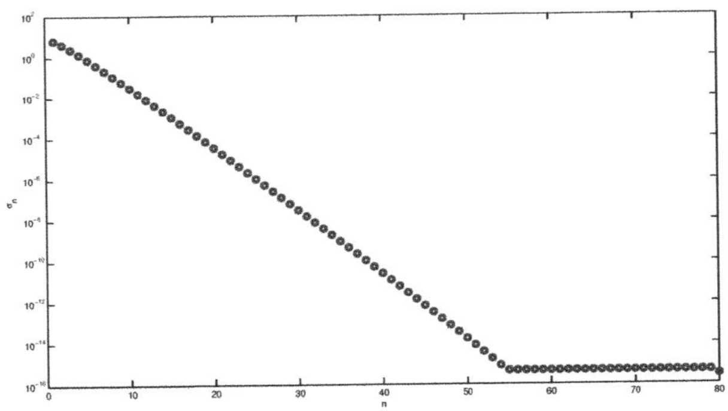

Figure 4.4. *Singular values for the operator related to gravimetric prospecting*

4.4. Applications of the SVD to least squares problems

4.4.1. *The matrix case*

The singular value decomposition provides the clearest solution to the least squares problems. We will use it in this section both to get a better theoretical understanding, in particular of perturbation issues, and we will also see that it constitutes a practical solution method.

1 http://www.mathworks.com/matlabcentral/fileexchange/52-regtools.

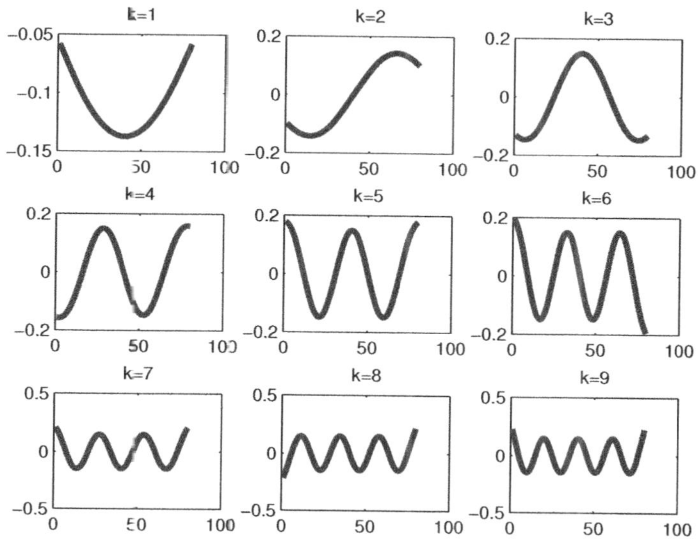

Figure 4.5. *Singular vectors for the operator related to gravimetric prospecting*

In this section, we no longer assume that A is of rank n since one of the strengths of the SVD is precisely to allow the general case to be addressed. When A is of rank $r < n$, we look for the minimal norm solution.

Let us introduce the SVD of A (equation [4.5]) in the least squares problem.

$$\|Ax - z\|_2^2 = \left\|U\Sigma V^T x - z\right\|_2^2 = \left\|\Sigma(V^T x) - U^T z\right\|_2^2, \qquad [4.14]$$

since U is orthogonal. We shall denote $w = U^T z$ (with $\|w\|_2 = \|z\|_2$), and choose $y = V^T x$ as the new unknown (with still $\|y\|_2 = \|x\|_2$, since V is orthogonal, which will be significant for computing the minimal norm solution). Since Σ is diagonal, this problem is decoupled, and is solved componentwise in the bases $(u_1, \ldots, u_m)$ and $(v_1, \ldots, v_n)$.

We thus have

$$\|Ax - z\|_2^2 = \|\Sigma y - w\|_2^2 = \sum_{i=1}^{r} |\sigma_i y_i - w_i|^2 . \qquad [4.15]$$

All the solutions of problem [4.2] are thus obtained by defining:

$$y_i = \begin{cases} w_i/\sigma_i, & \text{for } i = 1, \ldots, r \\ \text{unspecified} & \text{for } i = r+1, \ldots, n. \end{cases} \qquad [4.16]$$

We recover the results of section 4.1.1: in the case where A is of full rank, there is effectively a unique solution, and otherwise, the solution is defined up to the addition of a kernel element of A. In the latter case, the solution norm is $\sum_{i=1}^{r} |w_i/\sigma_i|^2 + \|\text{kernel element}\|_2^2$. The minimal norm solution is therefore the solution that has no component in $\operatorname{Ker} A$. Let us write the result in a more intrinsic form:

THEOREM 4.5.– The minimal norm solution of problem [4.2] is given by:

$$x = \sum_{i=1}^{r} (z, u_i)/\sigma_i \, v_i. \qquad [4.17]$$

PROOF.– We have already done most of the proof. By definition, $y = V^T x$ is inverted by $x = Vy$, whose components are $(x, v_i) = (Vy, v_i) = (y, V^T v_i) = y_i$. By using [4.16], we have

$$x = \sum_{i=1}^{r} (x, v_i)\, v_i = \sum_{i=1}^{r} y_i \, v_i = \sum_{i=1}^{r} (z, u_i)/\sigma_i \, v_i.$$

□

This result allows us to give a few first indications about the sensitivity of the solution x relatively to perturbations on the data z. To simplify, we shall consider the case where A is of rank n. Replace z by $z + \delta z$, and denote by $x + \delta x$ the solution of the perturbed problem. By linearity, it can be immediately deduced from [4.17] that

$$\delta x = \sum_{i=1}^{r} (\delta z, u_i)/\sigma_i \, v_i \qquad [4.18]$$

and thus that $\|\delta x\|_2^2 = \sum_{i=1}^{r} |\delta z|^2 / \sigma_i^2$.

In the absence of more precise information, the best that can be deduced from [4.18] is

$$\|\delta x\|_2 \leq \frac{\|\delta z\|_2}{\sigma_n},$$

since σ_n is the smallest singular value of A, which means that the error in the solution has been amplified by *the inverse of the smallest singular value* relatively to the error on the data. In the case where this smallest singular value is small, this magnification can become dramatic. We will see that this is the general case for the discretization of ill-posed problems, and this constitutes the main explanation for the instability found when solving these problems.

4.4.2. *The operator case*

As in the finite dimensional case, the singular value expansion of A allows for a complete analysis of the linear equations associated with the operator A such as [4.1] or [4.3].

THEOREM 4.6.– Let $z \in F$. Equation [4.1] has a solution in E if and only if $z \in \overline{\operatorname{Im} A}$ and if additionally:

$$\sum_{j=1}^{+\infty} \frac{|(z, f_j)|^2}{\sigma_j^2} < \infty. \qquad [4.19]$$

In this case, the set of solutions of [4.1] is given by

$$x = \sum_{j=1}^{+\infty} \frac{(z, f_j)}{\sigma_j} e_j + \operatorname{Ker} A \qquad [4.20]$$

PROOF.– We expand z on the Hilbert basis (f_j) : $z = \sum_{j=1}^{+\infty}(z, f_j)f_j$, and similarly for x on the Hilbert basis (e_j): $x = \sum_{j=1}^{+\infty} x_j e_j$. By applying [4.12], we get $Ax = \sum_{j=1}^{+\infty} \sigma_j x_j f_j$. It then suffices to identify the two expansions to obtain [4.20]. Condition [4.19] is simply obtained by expressing that the coefficients of series expansion of x must be square integrable. □

REMARK 4.5.– The resulting condition in [4.19] is called the *Picard condition*. It was obtained by this French mathematician at the beginning of the 20th Century. It expresses a restriction on the coefficients for an element to be in the image of A. These must tend toward 0 more quickly than required by the mere membership to $\overline{\operatorname{Im} A}$. Indeed, the sequence σ_j tends to 0, thus $|(z, f_j)|^2/\sigma_j^2 \leq |(z, f_j)|^2$ for j large enough. In particular, these coefficients should decrease more rapidly than the singular values. This condition (or its analogue in the finite dimensional case) can be checked numerically (see [HAN 90, HAN 98]).

REMARK 4.6.– The minimal norm solution of [4.1] is obviously obtained when the contribution of $\operatorname{Ker} A$ is zero (as a matter of fact, the series is orthogonal to $\operatorname{Ker} A$ since $f_j \in \operatorname{Im}$, for all j, and therefore the sum is in $\overline{\operatorname{Im} A}$).

EXAMPLE 4.6.– We continue example 4.4. According to Bessel–Parseval's identity, we get $\|u\|_{L^2}^2 = \sum_{k=1}^{\infty} |u_k|^2$. Here, the condition $u \in \operatorname{Im} A$ is defined by $u = \sum_{k=1}^{\infty} u_k f_k$, with $\sum_{k=1}^{\infty} k^2 |u_k|^2 < \infty$, which is equivalent to $u \in H^1(0,1)$.

EXAMPLE 4.7 (Gravimetric prospection).– We illustrate Picard's condition with the gravimetric prospection example 2.12.

We remain under the conditions of example 4.5. In addition to the matrix, we have a vector x_{ex} representing the solution. The right-hand side is calculated as $b_{\text{ex}} = Ax_{\text{ex}}$. We have computed the singular values of matrix A for $n = 80$, as well as the Fourier coefficients for the "exact" right-hand side b_{ex}, and for a noisy right-hand side, computed as $\tilde{b} = b_{\text{ex}} + 10^{-4}e$, where e is a random vector, whose elements follow a normal centered reduced distribution. Figure 4.6 illustrates the two situations, for a value of $n = 80$. We have represented, for the case of a second "exact" right-hand side member, the singular values (see Figure 4.4), the Fourier coefficients of b (coefficients in the base of the singular vectors) and the ratio of these two quantities:

– In the upper figure, the right-hand side has been calculated to be in the image of A, and therefore the Picard condition is satisfied. It can effectively observed that the Fourier coefficients of the right-hand side decrease faster than the singular values (at least up to the machine precision level).

– However, in the lower figure, the right-hand side is perturbed by white noise, and it can be seen that the Fourier coefficients decrease before stabilizing at a value between 10^{-5} and 10^{-4}, which corresponds to the noise level. The Picard condition will not be satisfied in this case.

REMARK 4.7.– By going back to the reasoning on stability in section 4.4.1 , the reasons that make the inverse problem ill posed can be recovered by means of the inversion formula [4.20] (at least in the case of a compact operator), and in particular the instability can be explained. Suppose that the component z_i of the second exact member is perturbed to $z_i' = z_i + \eta$. Through linearity, it can easily be seen that the only component of the solution x' that is modified is $x_i' = x_i + \eta/\sigma_i$. Now considering the norms, we obtain:

$$\|x' - x\|_E = \frac{\eta}{\sigma_i}. \qquad [4.21]$$

Since $\sigma_i \underset{i\to\infty}{\longrightarrow} 0$, it is no longer possible to uniformly bound the difference between x and x' as a function of $z - z'$. We thus find once again the conclusion that the solution of [4.1] does not depend continuously on the right-hand side.

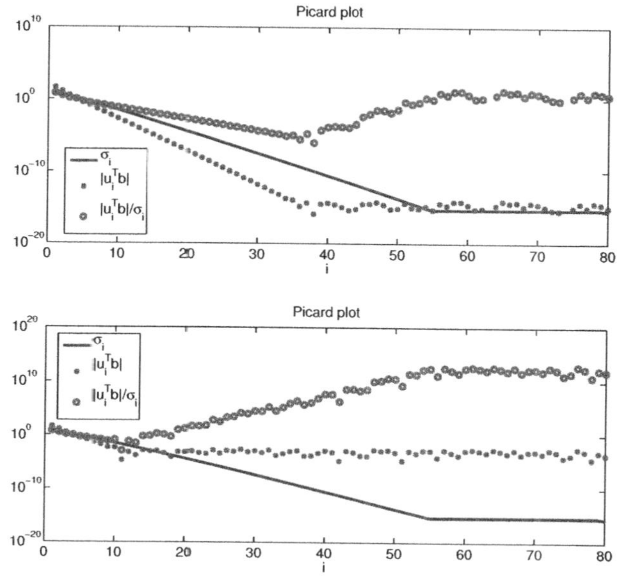

Figure 4.6. *Verification of the Picard condition for the gravimetric prospection example. Singular values, Fourier coefficients of the right-hand side, and ratio of the two. Upper image: non-noisy right-hand side, lower image: noisy right-hand side (noise level* 10^{-4}*). For a color version of the figure, see www.iste.co.uk/kern/inverse.zip*

4.5. Exercises

EXERCISE 4.1.– Orthogonal projections.

Throughout the whole exercise, we work in $\mathbf{R}^m$.

A *projector* is a matrix satisfying the property $P^2 = P$.

1) Show that any element in the space decomposes in a unique manner as a sum $x = x_1 + x_2$, with $x_1 \in \operatorname{Im} P$ and $x_2 \in \operatorname{Ker} P$.

Show that $I - P$ is also a projector, and that $\operatorname{Im}(I - P) = \operatorname{Ker} P$, $\operatorname{Ker}(I - P) = \operatorname{Im} P$.

2) In the case where $\operatorname{Ker} P$ and $\operatorname{Im} P$ are orthogonal, the projector is said to be *orthogonal*. Show that this definition is equivalent to $P^T = P$ (knowing that P is a projector).

Let F be a subspace of dimension n, and let $(q_1, \ldots, q_n)$ be an orthonormal basis of F. We denote by Q the matrix whose columns are the vectors $(q_1, \ldots, q_n)$. Show that the orthogonal projector on F has QQ^T for matrix

Application: Let $A = \begin{bmatrix} 1 & 0 \\ 0 & 1 \\ 1 & 0 \end{bmatrix}$. Determine the orthogonal projector on $\operatorname{Im} A$.

3) In the case of a least squares problem $\min_{x \in \mathbf{R}^n} \|Ax - b\|_2^2$, it is shown that the solution x satisfies $Ax = Pb$, where P is the orthogonal projector on $\operatorname{Im} A$.

EXERCISE 4.2.– SVD and QR Factorization

Let A be the matrix $A = \begin{bmatrix} \sqrt{2} & 0 \\ -1 & 1 \\ 1 & 1 \end{bmatrix}$.

1) Calculate the SVD of A.

2) Without any further calculation, derive the QR factorization of A.

EXERCISE 4.3.– Least squares and statistics.

In the standard linear model used in statistics, the hypothesis is made that the observations are perturbed by a random error and that the errors have zero mean, are independent, and all have the same variance σ^2. In other words, we get a relation of the form:

$$Ax = z = \hat{z} + e,$$

where the matrix $A \in \mathbf{R}^{n \times m}$ is fixed, and the vector e is a random vector, with

$$E(e) = 0; \qquad \operatorname{Var}(e) = \sigma^2 I.$$

The solution x of the least squares problem is thus also a random variable. We denote by $\hat{x}$ the (least squares) solution of $Ax = z$.

1) Show that

$$E(x) = \hat{x} = (A^T A)^{-1} A^T b, \quad \mathrm{Var}(x) = \sigma^2 (A^T A)^{-1}.$$

It is said that $\hat{x}$ is an *unbiased estimator* of x.

Hint: First show that if y is a random vector, and F a fixed matrix:

$$\mathrm{Var}(Fy) = F^T \,\mathrm{Var}(y) F$$

2) Show that, among unbiased estimators of x, $\hat{x}$ is the one that has the minimum variance (Gauss–Markov theorem).

EXERCISE 4.4.– Generalized least squares.

Let $A \in \mathbf{R}^{m \times n}$, $b \in \mathbf{R}^m$, and let $W \in \mathbf{R}^{m \times m}$ be a positive definite symmetric matrix . Consider the generalized problem:

$$\min_{x \in \mathbf{R}^n} (Ax - b)^T W^{-1} (Ax - b). \qquad [4.22]$$

1) In the particular case where $W = \mathrm{diag}(w_1, \dots, w_m)$, with $w_i > 0$, $i = 1, \dots, m$, show that if we let $D = \mathrm{diag}(\sqrt{w_1}, \dots, \sqrt{w_m})$, problem [4.22] is equivalent to

$$\min_{x \in \mathbf{R}^n} \left\| D^{-1}(Ax - b) \right\|_2^2 .$$

Explain why this corresponds to giving different weights to the various observations of the model.

2) Write the normal equations corresponding to this generalized problem.

Show that if we denote x_w the solution of the problem corresponding to a matrix W, and x the solution with $W = I$, we obtain

$$x_w - x = (A^T W^{-1} A)^{-1} A^T W^{-1} (b - Ax).$$

Conclude that if $b \in \mathrm{Im}\, A$, the two solutions are identical, but otherwise, the two solutions are different.

EXERCISE 4.5.– Singular values of matrices.

This exercise is from [TRE 97].

Determine the SVD of the following matrices (by hand, give an exact result).

$$\text{(a)} \begin{pmatrix} 3 & 0 \\ 0 & -2 \end{pmatrix}, \quad \text{(b)} \begin{pmatrix} 2 & 0 \\ 0 & 3 \end{pmatrix}, \quad \text{(c)} \begin{pmatrix} 0 & 2 \\ 0 & 0 \\ 0 & 0 \end{pmatrix}, \quad \text{(d)} \begin{pmatrix} 1 & 1 \\ 0 & 0 \end{pmatrix}$$

Give details of the calculations or the reasoning that allows to avoid them.

EXERCISE 4.6.– Singular values of an integral operator.

This exercise uses the notations and the results of exercise 3.1.

1) *Preliminary question*: Given the boundary problem

$$\begin{cases} -u''(t) = \lambda u(t), & t \in]0, 1[\\ u(0) = 0, \quad u'(1) + u(1) = 0. \end{cases}$$

Without solving the equation, show that $\lambda > 0$ (Hint: multiply the equation by u and integrate by parts).

By explicitly solving the equation, show that λ is a solution of a transcendental equation. Show that this equation has an infinite number of positive solution $(\lambda_n)_{n \in \mathbf{N}}$ (Hint: reason graphically).

2) Denote by (σ, u, v) a singular triplet for A ($Av = \sigma u, \;\; A^{\star}u = \sigma v$). It is shown that u and v are solutions of a differential system with constant coefficients that will be specified, as well as the boundary conditions on u and v.

3) Show that the singular values can be calculated from the numbers $(\lambda_n)_{n \in \mathbf{N}}$ found in the preliminary question.

EXERCISE 4.7.– Poisson operator.

Let A the integral operator on $L^2(-\pi, \pi)$ defined by

$$Au(t) = \int_{-\pi}^{\pi} \frac{1 - r^2}{1 - 2r\cos(t-s) + r^2} u(s)\, ds,$$

where $r < 1$ is a fixed parameter.

1) Verify the series expansion:

$$1 + \sum_{n=1}^{\infty} r^n e^{int} + \sum_{n=1}^{\infty} r^n e^{-int} = \frac{1 - r^2}{1 - 2r\cos(t-s) + r^2}. \qquad [4.23]$$

Conclude that

$$\int_{-\pi}^{+\pi} \frac{1 - r^2}{1 - 2r\cos(t-s) + r^2} \cos(ns)\, ds = r^n \cos(nt)$$

$$\int_{-\pi}^{+\pi} \frac{1 - r^2}{1 - 2r\cos(t-s) + r^2} \sin(ns)\, ds = r^n \sin(nt)$$

2) Deduce from [4.23] that

$$\left| \frac{1 - r^2}{1 - 2r\cos(t-s) + r^2} \right| \leq \frac{1+r}{1-r}$$

then that A is a compact operator of $L^2(-\pi, \pi)$ in itself.

3) Determine the adjoint of A.

4) It is recalled that the functions $(\cos(nt), \sin(nt))_{n \in \mathbf{N}}$ form a Hilbert basis $L^2(-\pi, \pi)$. Determine then the singular values and the singular vectors of A.

5

Regularization of Linear Inverse Problems

In this chapter, we will discuss regularization methods for linear inverse problems. Regularizing an ill-posed problem amounts to "replacing" it by another well-posed problem, such that the error is compensated by the gain in stability. Obviously, this requires that this replacement be quantified, which we will do. This chapter provides an introduction to the most common regularization methods, namely *Tikhonov's method* and *spectral truncation*. Obviously, it is not possible to reconstruct missing information, and regularization will lead to a loss of precision in the solution, which we will try to quantify according to the error from the data. We will see that it is possible to analyze Tikhonov's method in a very general (variational) context, but using singular value decompostion (SVD) makes it possible to bring into play some particular insights concering these methods.

The main difficulty in the application of a regularization method to a particular problem is the determination of the regularization parameter itself. We will discuss possible solutions. We will finish with a few words on iterative methods, by analyzing the simplest of them, the Landweber method.

In this chapter, A denotes a continuous linear operator from a Hilbert space E to a Hilbert space F, and we assume that the corresponding problem is *ill-posed*, that is to say that A is not invertible in $\mathcal{L}(E, F)$. As we discussed in Chapter 4, this can be due to the fact that A is not injective, but the most interesting case is that in which the image of A is not closed, which will always be the case if A is compact. In this case, A may or may not be bijective, but the inverse, if it exists, will not be continuous.

5.1. Tikhonov's method

5.1.1. *Presentation*

In order to solve the aforementioned instability, we will introduce *a priori* information.

We thus choose an *a priori* estimate $x_0 \in E$. For a quantity $\varepsilon > 0$ (the regularization parameter), we replace [4.2] by the *regularized problem*:

$$\min_{x \in E} \left\{ \frac{1}{2} \|Ax - \hat{z}\|_F^2 + \frac{\varepsilon^2}{2} \|x - x_0\|_E^2 \right\}. \qquad [5.1]$$

We will see that this problem admits a unique solution, which depends continuously on $\hat{z}$, and which converges, when $\varepsilon \to 0$, toward the solution of [4.2] that is closest to x_0. Obviously, if ε is chosen too small, [5.1] will be close to [4.2], therefore ill-posed, whereas if ε is too large, [5.1] is only used to force x to approximate x_0. The *optimal* choice of ε is thus delicate. We will return to this topic in section 5.3.

We start with the following result, which shows that [5.1] is still a least squares problem:

LEMMA 5.1.– Let $\tilde{A} = \begin{pmatrix} A \\ \varepsilon I \end{pmatrix}, \quad \tilde{z} = \begin{pmatrix} \hat{z} \\ \varepsilon x_o \end{pmatrix}, \quad \tilde{A} \in \mathcal{L}(E, F \times E).$

Then, [5.1] is equivalent to:

$$\min_{x \in E} \frac{1}{2} \|\tilde{A}x - \tilde{z}\|_{F \times E}^2. \qquad [5.2]$$

PROOF.– In effect, we simply calculate:

$$\tilde{A}x - \tilde{z} = \begin{pmatrix} Ax - \hat{z} \\ \varepsilon(x - x_0) \end{pmatrix}.$$

Next, it suffices to calculate the square of the norm. □

This result will easily allow us to show that [5.1] has a unique solution, but it is also used as basis for the numerical methods to solve [5.1] (see [ENG 96, HAN 98] and section 5.3.3 in this chapter). A consequence of lemma 5.1 and theorem 4.2 is the following formulation of [5.1].

PROPOSITION 5.1.– [5.1] is equivalent to

$$(A^*A + \varepsilon^2 I)x = A^*z + \varepsilon^2 x_0. \qquad [5.3]$$

This problem (and therefore [5.1]) admits a unique solution, which depends continuously on $\hat{z}$.

PROOF.– In effect, [5.3] is none other than the normal equation for [5.1]. It is obtained by observing that:

$$\tilde{A}^* = (A^*, \varepsilon I).$$

With regard to the existence and uniqueness of solutions to [5.3], it should be noted that:

$$((A^*A + \varepsilon^2 I)x, x) = \|Ax\|^2 + \varepsilon^2\|x\|^2 \geq \varepsilon^2\|x\|^2.$$

We then apply the Lax–Milgram lemma. According to theorem A3.2, the operator $A^*A + \varepsilon^2 I$, continuous and bijective, has a continuous inverse. In fact, we can obtain an explicit estimate by taking the inner product of equation [5.3] with x, which yields:

$$\|Ax\|^2 + \varepsilon^2\|x\|^2 \leq \|A^*\hat{z}\|\|x\| + \varepsilon^2\|x_0\|\|x\|,$$

that is to say:

$$\|x_\varepsilon\| \leq \frac{1}{\varepsilon^2}\|A^*\hat{z}\| + \|x_0\|. \qquad [5.4]$$

□

REMARK 5.1.– The estimate [5.4] "explodes" when $\varepsilon \to 0$. This is normal, since the solution $\hat{x}$ of [4.2] does not depend continuously on $\hat{z}$.

5.1.2. *Convergence*

We now want to address the question as to the extent that the Tikhonov method has correctly regularized the initial problem. To this end, it will be natural to consider the case that the data are noisy, since we then have no estimate about the error occurring in the solution. We will show that the Tikhonov method gives such an estimate, even if it is non-optimal, that is to say of a lower order than the error from the data.

We assume as known an *ideal* observation $\hat{z} \in \operatorname{Im} A$, and also a sequence of noisy data $z_n \in F$, $z_n \notin \operatorname{Im} A$, with $\delta_n = \|z_n - \hat{z}\|_F \underset{n\to\infty}{\longrightarrow} 0$. The quantity $\|z_n - \hat{z}\|_F \,/\, \|\hat{z}\|_F$ is the inverse of the *signal to noise ratio*. The underlying hypothesis in this section is that this quantity tends toward 0; in other words, it is possible to reduce the noise arbitrarily, which is obviously unrealistic.

Consider first the sequence of problems:

$$\text{Find } x_\varepsilon^n \in E \text{ achieving the minimum of } \frac{1}{2}\|Ax - z_n\|_F^2 + \frac{\varepsilon^2}{2}\|x - x_0\|_E^2, \qquad [5.5]$$

(x_ε^n exists and is unique according to proposition 5.1). Note that we do not, for now, seek to adjust the regularization parameter to the noise level.

To understand how the regularization method "works," we are going to estimate the error between the solution of the noisy problem and the exact solution. We will thus highlight a first manifestation of the *stability–accuracy compromise* that results from a regularization method. For simplicity, we will make the (regularity) hypothesis that $\hat{x} - x_0 \in \operatorname{Im} A^\star$. The general case is addressed by Baumeister [BAU 87].

PROPOSITION 5.2.– Suppose that $\hat{x} - x_0 \in \operatorname{Im} A^\star$, say $\hat{x} = A^\star w$. Then:

$$\forall n, \; \|x_\varepsilon^n - \hat{x}\|_E \leq \|A^\star\| \frac{\delta_n}{\sqrt{2}\varepsilon} + \frac{\varepsilon}{\sqrt{2}}\|w\|_F$$

PROOF.– Let us introduce the intermediate quantity x_ε as solution of the problem:

$$A^\star A x_\varepsilon + \epsilon^2 x_\varepsilon = A^\star \hat{z} + \varepsilon^2 x_0. \qquad [5.6]$$

The triangle inequality gives:

$$\|x_n - \hat{x}\| \leq \|x_n - x_\varepsilon\| + \|x_\varepsilon - \hat{x}\|. \qquad [5.7]$$

We will separately estimate each term of the right-hand side in this inequality. The first part corresponds to the error from the data, and is amplified by the ill-posed nature of the underlying problem, while the second is due to the approximation of the exact solution, and tends to 0 with ε.

By subtracting [5.6] from the normal equation associated to [5.5], we obtain:

$$\varepsilon^2(x_\varepsilon^n - x_\varepsilon) + A^\star A(x_\varepsilon^n - x_\varepsilon) = A^\star(z_n - \hat{z}),$$

then taking the inner product with $x_\varepsilon^n - x_\varepsilon$ gives:

$$\begin{aligned}\varepsilon^2\|x_\varepsilon^n - x_\varepsilon\|^2 + \|A(x_\varepsilon^n - x_\varepsilon)\|^2 &\leq \|z_n - \hat{z}\| \, \|A(x_\varepsilon^n - x_\varepsilon)\| \\ &\leq \frac{1}{2}\left(\|z_n - \hat{z}\|^2 + \|A(x_\varepsilon^n - x_\varepsilon)\|^2\right)\end{aligned}$$

(using the elementary inequality $ab \leq \frac{1}{2}(a^2 + b^2)$), so that:

$$\varepsilon \left\| x_\varepsilon^n - x_\varepsilon \right\| \leq \frac{1}{\sqrt{2}} \delta_n. \qquad [5.8]$$

which is a bound for the first part of [5.7].

For the second part, we first write:

$$\left\| Ax_\varepsilon - A\hat{x} \right\|^2 = (Ax_\varepsilon - A\hat{x},\ Ax_\varepsilon - A\hat{x}) = (A^\star Ax_\varepsilon - A^\star A\hat{x},\ x_\varepsilon - \hat{x})$$

by the definition of $A^\star$. Using the normal equations satisfied, respectively, by x_ε and by $\hat{x}$, we obtain:

$$\begin{aligned} \left\| Ax_\varepsilon - A\hat{x} \right\|^2 &= (-\varepsilon^2 x_\varepsilon - \varepsilon^2 x_0,\ x_\varepsilon - \hat{x}) = -\varepsilon^2 (x_\varepsilon - \hat{x} + \hat{x} - x_0,\ x_\varepsilon - \hat{x}) \\ &= -\varepsilon^2 \left\| x_\varepsilon - \hat{x} \right\|_E + \varepsilon^2 (x_0 - \hat{x}.\ x_\varepsilon - \hat{x}) \end{aligned} \qquad [5.9]$$

The regularity hypothesis on $\hat{x}$ allows us to rewrite the second inner product as:

$$(x_0 - \hat{x},\ x_\varepsilon - \hat{x}) = (A^\star w,\ x_\varepsilon - \hat{x}) = (w, A(\ x_\varepsilon - \hat{x})),$$

and this can be bounded by:

$$|(x_0 - \hat{x},\ x_\varepsilon - \hat{x})| \leq \|w\|_F^2 \left\| Ax_\varepsilon - A\hat{x} \right\|^2 \leq \frac{1}{2} \|w\|_F^2 + \frac{1}{2} \left\| Ax_\varepsilon - A\hat{x} \right\|^2 \qquad [5.10]$$

by the Cauchy–Schwarz inequality and the elementary inequality used in the first part.

By substituting [5.10] in [5.9], we obtain the bound:

$$\frac{1}{2} \left\| A(x_\varepsilon - \hat{x}) \right\|^2 + \varepsilon^2 \left\| x_\varepsilon - \hat{x} \right\|^2 \leq \frac{\varepsilon^4}{2}, \|w\|^2$$

and in particular

$$\left\| x_\varepsilon - \hat{x} \right\| \leq \frac{\varepsilon}{\sqrt{2}} \|w\| . \qquad [5.11]$$

By gathering [5.8] and [5.11], we obtain the estimate of the theorem. □

This result shows that, as we have pointed out, the error is composed of two terms:

– a first term due to errors from data, multiplied by a *condition number*, which tends toward infinity when $\varepsilon \to 0$;

– a second term due to the approximation of the exact solution, which tends toward 0 with ε.

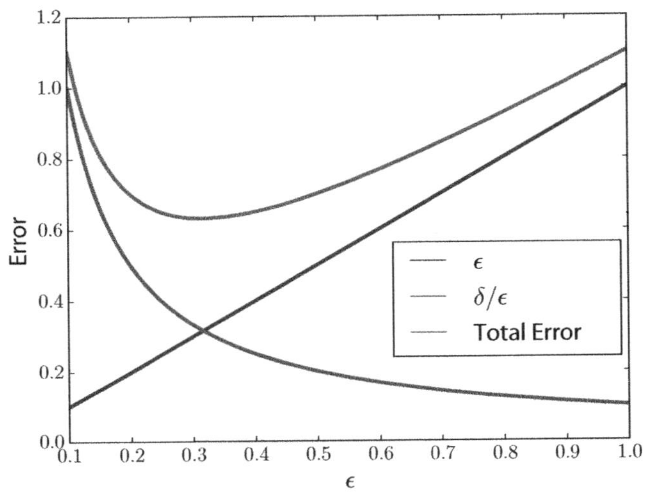

Figure 5.1. *Behavior of the total error. For a color version of the figure, see www.iste.co.uk/kern/inverse.zip*

We thus can clearly see the need to adjust the regularization parameter to the noise level present in the data. Such a *regularization strategy* may be designed in two ways:

– if an estimate of the noise level is available, it is possible to make use of it to choose ε so as to obtain convergence of x_ε^n toward $\hat{x}$. This is what we do in the following theorem. Such a strategy is called an *a priori* regularization strategy. It assumes that we know how to estimate the noise existing in the data, which may not always be possible;

– the other strategy, called *a posteriori*, consists of estimating during the computation a suitable value of the parameter, by using only the available data. Such strategies exist (see [ENG 96, HAN 98]). We will see a few examples in section 5.3.1.

In proposition 5.2, we have allowed ε to tend to 0 regardless of δ and we have seen that such a strategy (or rather the lack thereof) does not allow the convergence of the regularized solution toward the "real" solution. We are therefore going to adjust the regularization parameter to the noise level. To this end, we modify problem [5.5]:

$$\text{find } x_n \in E \text{ achieving the minimum of } \frac{1}{2}\|Ax - z_n\|_F^2 + \frac{\varepsilon_n^2}{2}\|x - x_0\|_F^2, \qquad [5.12]$$

where we are looking for a way to choose the sequence ε_n to ensure the convergence of x_n toward *a* solution of [4.2]. It turns out that this solution will always be the closest solution to x_0. This result is a consequence of the following lemma.

LEMMA 5.2.– For $x \in E$, note the (unique) decomposition:

$$x = x^K + x^I, \quad x^K \in \operatorname{Ker} A, \quad x^I \in \operatorname{Ker} A^\perp = \overline{\operatorname{Im} A^*}$$

The solution of problem [5.12] satisfies:

$$\forall n, \quad x_n^K = x_0^K.$$

PROOF.– We project orthogonally equation [5.3] upon $\operatorname{Ker} A$:

– $A^*Ax_n \in \operatorname{Im} A^* \subset (\operatorname{Ker} A)^\perp$

– $A^*z_n \in \operatorname{Im} A^* \subset (\operatorname{Ker} A)^\perp$

where $\varepsilon^2 x_n^K = \varepsilon^2 x_0^K$. □

We can now state the convergence theorem, whose elementary proof is borrowed from Chavent [CHA 96].

THEOREM 5.1.– Let $\hat{z} \in \operatorname{Im} A$. Suppose that $\delta_n = \|z_n - \hat{z}\| \underset{n\to\infty}{\longrightarrow} 0, \quad \varepsilon_n \underset{n\to\infty}{\longrightarrow} 0$. Then:

1) $\|Ax_n - \hat{z}\|_F \underset{n\to\infty}{\longrightarrow} 0$

2) If $\delta_n/\varepsilon_n \underset{n\to\infty}{\longrightarrow} 0$, then $\|Ax_n - \hat{z}\|_F = O(\varepsilon_n)$ and $x_n \underset{n\to\infty}{\longrightarrow} \hat{x}$, where $\hat{x}$ is the solution of [4.2] closest to x_0.

3) If in addition $\hat{x} - x_0 \in \operatorname{Im} A^*$ (a regularity hypothesis), and if $\delta_n/\varepsilon_n^2 \underset{n\to\infty}{\longrightarrow} 0$ then $\|Ax_n - \hat{z}\|_F = O(\varepsilon_n^2)$ and $\|x_n - \hat{x}\|_E = O(\varepsilon_n)$

PROOF.– Denote by $\hat{x}$ the solution [4.3] that minimizes $\|x - x_0\|_E$ ($\hat{x}$ is well defined due to a variant of corollary 4.1).

1) By definition of x_n, we have:

$$\|Ax_n - z_n\|_F^2 + \varepsilon_n^2 \|x_n - x_0\|_E^2 \leq \|A\hat{x} - z_n\|_E^2 + \varepsilon_n^2 \|\hat{x} - x_0\|_E^2 ,$$

where (by adding and subtracting a positive quantity):

$$\begin{aligned} \|Ax_n - \hat{z}\|_F^2 + \varepsilon_n^2 \|x_n - \hat{x}\|_E^2 \leq {} & \|Ax_n - z\|_F^2 + \varepsilon_n^2 \|x_n - \hat{x}\|_E^2 \\ & + \|A\hat{x} - z_n\|_F^2 + \varepsilon_n^2 \|\hat{x} - x_0\|_E^2 \\ & - \|Ax_n - z_n\|_F^2 - \varepsilon^2 \|x_n - x_0\|_E^2 . \end{aligned}$$

Using the identities:

$$Ax_n - z_n = (Ax_n - \hat{z}) + (\hat{z} - z_n) \quad \text{and} \quad x_n - x_0 = (x_n - \hat{x}) + (\hat{x} - x_0)$$

we obtain:

$$\begin{aligned} \|Ax_n - \hat{z}\|_F^2 + \varepsilon_n^2 \|x_n - \hat{x}\|_E^2 & \leq -2(Ax_n - \hat{z}, \hat{z} - z_n) - 2\varepsilon_n^2(x_n - \hat{x}, \hat{x} - x_0) \\ & \leq 2\delta_n \|A\hat{x}_n - z\|_F + 2\varepsilon_n^2 \|x_n - \hat{x}\|_E \|\hat{x} - x_0\|_E \end{aligned} \quad [5.13]$$

by the Cauchy–Schwarz inequality. If we note, in the previous inequality:

$$\begin{aligned} a &= \|Ax_n - \hat{z}\|_F , & b &= \varepsilon_n \|x_n - \hat{x}\|_E , \\ \alpha &= \delta_n, & \beta &= \varepsilon_n \|\hat{x} - x_0\|_E , \end{aligned}$$

this inequality is rewritten:

$$a^2 + b^2 \leq 2a\alpha + 2b\beta$$

and we derive thereof:

$$(a - \alpha)^2 + (b - \beta)^2 \leq \alpha^2 + \beta^2 \leq (\alpha + \beta)^2,$$

then

$$\begin{cases} a - \alpha \leq \alpha + \beta \implies a \leq 2\alpha + \beta, \\ b - \beta \leq \alpha + \beta \implies b \leq \alpha + 2\beta, \end{cases}$$

that is to say finally:

$$\begin{cases} \|Ax_n - \hat{z}\|_F \leq 2\delta_n + \varepsilon_n \|\hat{x} - x_0\|_E \\ \|x_n - \hat{x}\|_E \leq \dfrac{\delta_n}{\varepsilon_n} + 2\|\hat{x} - x_0\|_E . \end{cases} \quad [5.14]$$

We therefore obtain the convergence of the observations under the only hypothesis that $\hat{z} \in \operatorname{Im} A$. However, this is normal since at this point we have not yet linked ε_n to δ_n, we cannot conclude as to the convergence of x_n toward $\hat{x}$.

2) In addition, we assume that $\delta_n/\varepsilon_n \underset{n\to\infty}{\longrightarrow} 0$. Let then $\eta > 0$ be fixed. The definition of $\hat{x}$ implies that $\hat{x} - x_0 \in \operatorname{Ker} A^\perp = \overline{\operatorname{Im} A^*}$, thus:

$$\exists w \in F, \quad \|\hat{x} - x_0 - A^* w\| \leq \eta.$$

We go back to [5.13], noting that $(x_n - \hat{x}, A^* w) = (A(x_n - \hat{x}), w) = (Ax_n - \hat{z}, w)$.

$$\begin{aligned} \|Ax_n - \hat{z}\|_F^2 + \varepsilon_n^2 \|x_n - \hat{x}\|_E^2 &\leq -2(A\hat{x}_n - \hat{z},\, \hat{z} - z_n) - 2\varepsilon_n^2(\hat{x}_n - x_0,\, \hat{x} - x_0) \\ &\quad + 2\varepsilon_n^2(Ax_n - \hat{z},\, w) - 2\varepsilon_n^2(x_n - \hat{x},\, A^* w) \\ &\leq 2\|Ax_n - \hat{z}\|_F (\delta_n + \varepsilon_n^2 \|w\|_F) + 2\varepsilon_n^2 \|x_n - \hat{x}\|_E \eta \end{aligned}$$

As previously, we derive thereof:

$$\begin{cases} \|Ax_n - \hat{z}\|_F \leq 2\delta_n + 2\varepsilon_n^2 \|w\|_F + \varepsilon_n \eta = \varepsilon_n \left(2\dfrac{\delta_n}{\varepsilon_n} + 2\varepsilon_n \|w\|_F + \eta \right), \\ \|x_n - \hat{x}\|_E \leq \dfrac{\delta_n}{\varepsilon_n} + \varepsilon_n \|w\|_F + 2\eta \end{cases} \quad [5.15]$$

and, since we have assumed that $\dfrac{\delta_n}{\varepsilon_n} \to 0$, the two quantities above may be bounded by $3\varepsilon_n \eta$ (respectively 3η) for n large enough, and since η is arbitrary, this proves the convergence of x_n to $\hat{x}$.

3) Finally, if we assume that $\hat{x} - x_0 \in \operatorname{Im} A^*$, we can set $\eta = 0$ in the inequalities [5.15].

In addition, assuming then $\dfrac{\delta_n}{\varepsilon_n^2} \underset{n\to\infty}{\longrightarrow} 0$, we obtain:

$$\begin{cases} \|Ax_n - \hat{z}\|_F \leq \varepsilon_n^2 \left(\dfrac{\delta n}{\varepsilon_n^2} + 2\|w\|_F \right) = O(\varepsilon_n^2), \\ \|x_n - \hat{x}\|_E \leq \varepsilon_n \left(\dfrac{\delta n}{\varepsilon_n^2} + \|w\|_F \right) = O(\varepsilon_n). \end{cases}$$

□

This theorem once again highlights the stability–accuracy compromise characteristic of ill-posed problems. It expresses that the sequence ε_n must tend toward 0 *less quickly* than the noise level if the convergence of the regularized solutions is to be obtained. Furthermore, when the solution becomes more regular, the sequence ε_n must converge to 0 more slowly. Concerning the error in the solution, it is of the order of $\varepsilon_n = O(\sqrt{\delta_n})$, which is therefore greater than the noise level. This thus results in a loss of accuracy, obviously due to instability. As we mentioned above, Tikhonov's method is not an optimal regularization method. For a more detailed discussion of this topic and for the analysis of another method, see [ENG 96].

REMARK 5.2.–

1) The hypothesis $\hat{x} - x_0 \in \operatorname{Im} A^*$ typically corresponds to a regularity result for $\hat{x}$, the solution of [4.3].

2) This theorem has an essentially theoretical interest. In effect, in order to apply it, it would be necessary to know the sequence δ_n, that is the *noise level* contained in the data, which is difficult in practice. The choice of ε_n given by theorem 5.1 is what we called above an *a priori* choice.

EXAMPLE 5.1 (Canonical injection).– We apply the Tikhonov method to the *canonical injection* operator of example 4.1. The original problem is: given $\hat{z} \in L^2(0,1)$, find $u \in H_0^1(0,1)$ such that $u = \hat{z}$. Since we can identify $u \in H_0^1(0,1)$ to the couple (u, u'), this problem amounts to finding u' from a measurement of u.

According to equation [4.4], the regularized solution u_ε is the solution of the following elliptic problem:

$$\begin{cases} -\varepsilon u'' + u = \hat{z} \\ u(0) = u(1) = 0. \end{cases} \qquad [5.16]$$

A computation with Maple gives:

$$u(t) = \frac{\sinh(t/\varepsilon)}{\sinh(1/\varepsilon)} \int_0^1 \frac{1}{\varepsilon} \sinh\left(\frac{1-s}{\varepsilon}\right) f(s)\, ds - \int_0^T \frac{1}{\varepsilon} \sinh\left(\frac{t-s}{\varepsilon}\right) f(s)\, ds.$$

As expected, u depends continuously on ε, but the passage to the limit is singular.

EXAMPLE 5.2 (Gravimetric prospecting).– We now illustrate the behavior of Tikhonov's method on our guiding example. To this end, we use the `tikhonov` function from the *Regularization Toolbox*. Note that the implementation of the method is based on the singular value decomposition of the matrix A, so as to make it simpler, but this has no impact on the results. In Figure 5.2, it can effectively be observed that the small values of the regularization parameter give a solution that remains still oscillatory (left figure, $\varepsilon = 10^{-5}$), while too much regularization "squeezes" the solution ($\varepsilon = 10$). The figure on the right shows that values between 10^{-3} and 1 lead to a suitable solution.

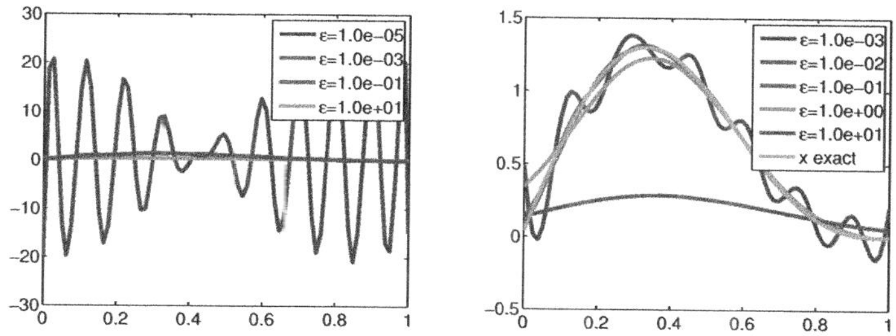

Figure 5.2. *Application of Tikhonov's regularization to the geomagnetic prospecting problem. For a color version of the figure, see www.iste.co.uk/kern/inverse.zip*

We will return to the sensitive issue of the choice of regularization parameters in section 5.3.1.

5.1.3. *The L-curve*

This is a tool for the analysis of regularized problem solutions, when the regularization parameter is varied. It is initially based on a visual inspection of the parametrized curve $(\|Ax^\varepsilon - z\|, \|x^\varepsilon\|)$ with respect to ε. The usefulness of this curve to analyze least squares problems was noticed by Lawson and Hanson [LAW 95]. Its importance in the study of inverse problems has been emphasized and its properties have been studied in the works of Hansen (see [HAN 92], or a detailed analysis in [HAN 98]). We will follow here the observations of Hansen [HAN 10].

We will need finer properties of the dependency of the solution, and of the residual with regard to the regularization parameter.

PROPOSITION 5.3.– The solution x^ε depends continuously on ε. The function $\varepsilon \mapsto \|x^\varepsilon\|_E$ is decreasing and tends toward 0 when $\varepsilon \to \infty$.

The function $\varepsilon \mapsto \|Ax^\varepsilon - z\|_F$ is increasing, and $\lim_{\varepsilon \to 0} Ax^\varepsilon = z$.

PROOF.– According to the definition of x^ε, we obtain:

$$\varepsilon^2 \|x^\varepsilon\|_E^2 \le \|Ax^\varepsilon - z\|_F + \varepsilon^2 \|x^\varepsilon\|_E^2 \le \|z\|_F^2$$

(by taking $x = 0$ in the function to minimize), that is

$$\|x^\varepsilon\|_E \le \sqrt{\|z\|_F/\varepsilon}.$$

For the remainder of the proof, we establish first the equality:

$$\varepsilon^2 \|x^\varepsilon - x^\eta\|^2 + \|A(x^\varepsilon - x^\eta)\|^2 = (\eta^2 - \varepsilon^2)(x^\eta, x^\varepsilon - x^\eta). \qquad [5.17]$$

This identity is obtained by subtracting the normal equations for x^ε and x^η:

$$\varepsilon^2(x^\varepsilon - x^\eta) + A^*A(x^\varepsilon - x^\eta) + (\varepsilon^2 - \eta^2)x^\eta = 0$$

and by taking the inner product with $(x^\varepsilon - x^\eta)$.

It follows from [5.17] that

$$\varepsilon^2 \|x^\varepsilon - x^\eta\|_E^2 \le \left|(\eta^2 - \varepsilon^2)\right| |(x^\eta, x^\varepsilon - x^\eta)| \le \left|(\eta^2 - \varepsilon^2)\right| \|x^\eta\|_E \|x^\varepsilon - x^\eta\|_E,$$

that is to say

$$\varepsilon^2 \|x^\varepsilon - x^\eta\|_E \le \left|(\eta^2 - \varepsilon^2)\right| \|x^\eta\|_E \le \left|(\eta^2 - \varepsilon^2)\right| \frac{\|z\|_F}{\varepsilon},$$

which shows the continuity of the function $\varepsilon \to x^\varepsilon$.

Let now $\eta > \varepsilon$. It follows from [5.17] that $(x^\eta, x^\varepsilon - x^\eta)$ is positive. As a consequence,

$$\|x^\eta\|_E^2 \le (x^\eta, x^\varepsilon) \le \|x^\eta\|_E \|x^\varepsilon\|_E,$$

so that the function $\varepsilon \to \|x^\varepsilon\|_E$ is indeed decreasing.

Let now $\eta \le \varepsilon$. We take the inner product of the normal equation for x^η by $x^\varepsilon - x^\eta$, to obtain:

$$\eta(x^\eta, x^\varepsilon - x^\eta) + (Ax^\eta - z, A(x^\varepsilon - x^\eta)) = 0.$$

By using [5.17], we can see that this time $(x^\eta, x^\varepsilon - x^\eta) \le 0$, and thus $0 \le (Ax^\eta - z, A(x^\varepsilon - x^\eta)) = (Ax^\eta - z, Ax^\varepsilon - z) - \|Ax^\eta - z\|_F^2$, and the Cauchy–Schwarz inequality shows that the application $\varepsilon \to \|Ax^\varepsilon - z\|_F$ is increasing.

Finally, let $\alpha > 0$. Since the image of A is dense in F, there exists $x \in E$ such that $\|Ax - z\|_F \le \alpha^2/2$. Take ε_0 such that $\varepsilon_0^2 \|x\|_E^2 \le \alpha^2/2$. Then:

$$\|Ax^\varepsilon - z\|_F^2 \le \|Ax^\varepsilon - z\|_F^2 + \varepsilon^2 \|x^\varepsilon\|_E^2 \le \|Ax - z\|_F^2 + \varepsilon^2 \|x\|_E^2 \le \alpha^2,$$

so that $\|Ax^\varepsilon - z\|_F^2 \le \alpha$, for $\varepsilon \le \varepsilon_0$. □

It can also be shown that the L-curve is convex. A finer analysis (see [HAN 93b]) indicates that the curve gets its name from the fact that it comprises two distinct parts:

– an essentially vertical part, for the small regularization parameter values. When this parameter decreases, the norm of the solution increases, without the residual being reduced. We are therefore moving vertically on the curve. This corresponds to an insufficient regularization;

– an essentially horizontal section, for large values of the regularization parameter. If the regularization parameter increases, the residual decreases (that is to say the quality of the solution is no longer improved), but the size of the solution eventually decreases.

This behavior is easier to visualize if the curve is represented in double logarithmic scale. Figure 5.3 shows an example of this curve in the case of our gravimetric prospection example. The regularization parameter varies between 10^{-6} and 10 (we have seen in Figure 5.2 that 10^{-6} is "too small" and that 10 is "too large"). We can see that the curve has effectively the expected shape, with a vertical branch, and a horizontal branch.

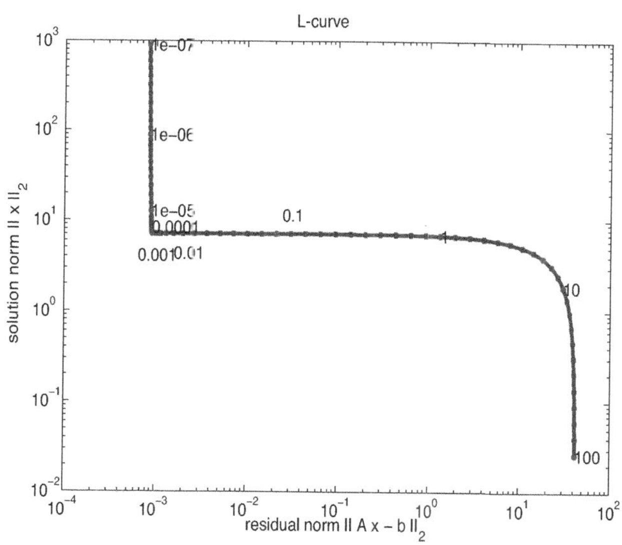

Figure 5.3. *L-curve example for the gravimetric prospecting example*

5.2. Applications of the SVE

We assume in this section that the operator A is compact. According to theorem 4.4, it admits a singular values expansion. We use the notation of section 4.3.

5.2.1. *Singular value expansion and Tikhonov's method*

We are now going to use the information provided by the SVE to regularize our problem. To this end, we reinterpret the Tikhonov method.

PROPOSITION 5.4.– Let $z \in F$. The solution of equation [5.3] can be expanded on the Hilbert basis (e_j) as follows:

$$x = \sum_{j=1}^{+\infty} \frac{\sigma_j}{\sigma_j^2 + \varepsilon^2}(z, f_j)e_j \qquad [5.18]$$

PROOF.– As for theorem 4.6, we expand the right-hand side and the solution according to their respective Hilbert bases, then we use the equations of [4.12] to obtain:

$$\sum_{j=1}^{+\infty} x_j(\sigma_j^2 + \varepsilon^2)f_j = \sum_{j=1}^{+\infty} \sigma_j(z, f_j)f_j.$$

It suffices then to identify the coefficients. □

REMARK 5.3.– Unlike the sequence $(1/\sigma_j)$, the sequence $\sigma_j/\sigma_j^2+\varepsilon^2$ remains bounded when $j \to \infty$, for a *fixed* $\varepsilon > 0$. We find thus once more the fact that the regularized problem is well-posed.

We see that the SVE results in a simple interpretation of Tikhonov's method. We have replaced the function $\sigma \mapsto 1/\sigma$, unbounded in the neighborhood of 0, by the function $\sigma \mapsto \sigma/\sigma^2 + \varepsilon^2$, which remains bounded. Naturally, the bound in question tends to infinity as ε goes to 0 (it is equal to $1/2\varepsilon$ for $\sigma = \varepsilon$). We illustrate this result in Figure 5.4.

REMARK 5.4.– If we go back to the calculation of remark 4.3, we obtain now:

$$\|x' - x\|_E = \eta \frac{\sigma_i}{\sigma_i^2 + \varepsilon^2} \leq \frac{\eta}{2\varepsilon} = \frac{1}{2\varepsilon}\|z - z'\|. \qquad [5.19]$$

This time a bound independent of the singular values of A is obtained.

REMARK 5.5.– Formula [5.18] can also be rewritten in the form:

$$x = \sum_{j=1}^{+\infty} \frac{\sigma_j^2}{\sigma_j^2 + \varepsilon^2}\frac{(z, f_j)}{\sigma_j}e_j, \qquad [5.20]$$

the coefficients $\frac{\sigma_j^2}{\sigma_j^2 + \varepsilon^2}$ are called the *filter factors* of Tikhonov's method. They are obviously between 0 and 1. We can plot (Figure 5.5) the analog of Figure 5.4 for this representation. It can be observed that the filter factors corresponding to the large singular values are only slightly damped, which is desirable since these singular values are the most representative, while those that correspond to small singular values (more specifically those singular values that satisfy $\sigma_j < \epsilon$) are effectively filtered.

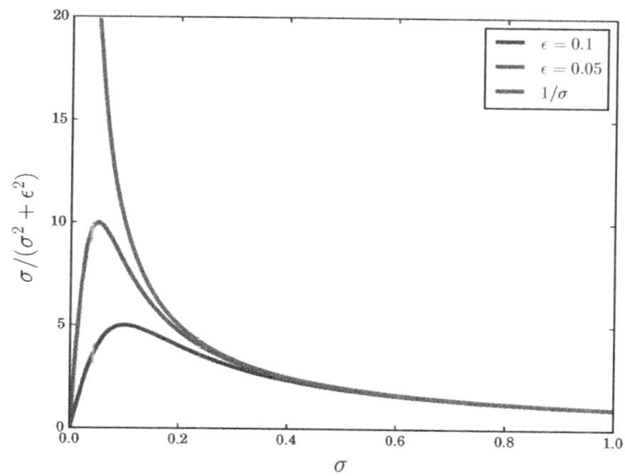

Figure 5.4. *Evolution of the SVE filter function with ϵ. For a color version of the figure, see www.iste.co.uk/kern/inverse.zip*

5.2.2. Regularization by truncated SVE

If the SVE of the operator A is known, an alternative regularization method can be proposed, called spectral truncation. As its name implies, this method consists of truncating expansion [4.11] to a certain rank. This rank fulfills the role of the regularization parameter ε in Tikhonov's method.

For $n > 0$, let

$$x_n = \sum_{j=1}^{n} \frac{(z, f_j)}{\sigma_j} e_j. \qquad [5.21]$$

Since the series converges, it is clear that $x_n \underset{j \to \infty}{\longrightarrow} \hat{x}$ (where $\hat{x}$ is the minimal norm solution of [4.2]). The truncation method converges thus with noiseless data. Let us now examine the case of noisy data, that is we replace $\hat{z}$ by z_δ, with $\|\hat{z} - z_\delta\| = \delta$.

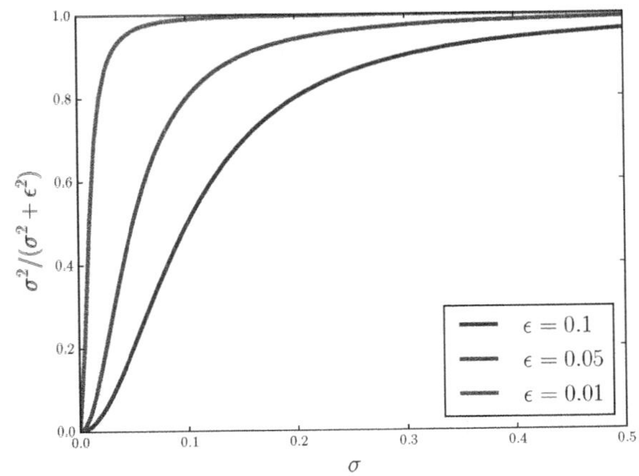

Figure 5.5. *Evolution of the SVD filter factors with ϵ. For a color version of the figure, see www.iste.co.uk/kern/inverse.zip*

Denoting

$$x_n^\delta = \sum_{j=1}^{n} \frac{(z^\delta, f_j)}{\sigma_j} e_j,$$

which is what can actually be computed, we aim to estimate the difference $x_n - x_n^\delta$. By definition

$$x_n - x_n^\delta = \sum_{j=1}^{n} \frac{(z - z^\delta, f_j)}{\sigma_j} e_j,$$

and thus, since the system e_j is orthonormal,

$$\left\|x_n - x_n^\delta\right\|^2 = \sum_{j=1}^{n} \frac{1}{\sigma_j^2} \left|(z - z^\delta, f_j)\right|^2 \le \frac{1}{\sigma_n^2} \sum_{j=1}^{n} \left|(z - z^\delta, f_j)\right|^2 \le \frac{1}{\sigma_n^2} \delta^2. \qquad [5.22]$$

With regard to the total error, we write:

$$\left\|\hat{x} - x_n^\delta\right\| \le \left\|\hat{x} - x_n\right\| + \left\|x_n - x_n^\delta\right\| \le \left\|\hat{x} - x_n\right\| + \frac{\delta}{\sigma_n}. \qquad [5.23]$$

The first term tends toward 0, and in order for the sum to tend toward 0, it is more important that n be chosen according to δ such that if, for example, we consider a sequence $z_n \underset{n\to\infty}{\longrightarrow} \hat{z}$, with $\|\hat{z} - z_n\| = \delta_n$, then $\dfrac{\delta_n}{\sigma_n} \underset{n\to\infty}{\longrightarrow} 0$.

To obtain a convergence rate, we must here also make a regularity hypothesis. Therefore, we will assume as before that $\hat{x} \in \operatorname{Im} A^\star$, say $\hat{x} = A^\star w,\ w \in F$. In this case, using [4.12]:

$$\begin{aligned} x_n - \hat{x} &= \sum_{j=1}^{n} \frac{(z, f_j)}{\sigma_j} e_j - \sum_{j=1}^{+\infty} \sigma_j (w, f_j) e_j \\ &= \sum_{j=1}^{n} \left[\frac{(z, f_j)}{\sigma_j} - \sigma_j (w, f_j) \right] e_j + \sum_{j=n+1}^{+\infty} \sigma_j (w, f_j) e_j, \end{aligned}$$

and thus (the singular values form a decreasing sequence):

$$\|x_n - \hat{x}\|_E \leq \sigma_{n+1} \|w\|_F .$$

Equation [5.23] becomes then:

$$\left\|\hat{x} - x_n^\delta\right\| \leq \sigma_{n+1} \|w\|_F + \frac{\delta_n}{\sigma_n}. \qquad [5.24]$$

A simple way to ensure convergence is, for example, to choose n so that $\delta_n/\sigma_n = C\sigma_{n+1}$ (where C is a constant). Inequality [5.24] then gives:

$$\left\|\hat{x} - x_n^\delta\right\| \leq C'\sigma_{n-1} = O(\sqrt{\delta_n}).$$

EXAMPLE 5.3 (Gravimetric prospecting).– Figure 5.6 illustrates the behavior of the regularization by spectral truncation on the example of the gravimetric prospecting example.

We can see that, as for Tikhonov's method, the choice of the regularization parameter (here n) is both important and delicate. If n is too large, the approximation will look excessively like the unstable one given by formula [4.20], and the influence of small singular values will result in instability. If n is too small, the obtained solution will only have little in common with the actual solution. In this discussion, it is actually important to take into account the more or less high speed with which the singular values tend toward 0. The faster this decrease, the smaller n must be, that is to say that only a small number of components of the solution can be recovered.

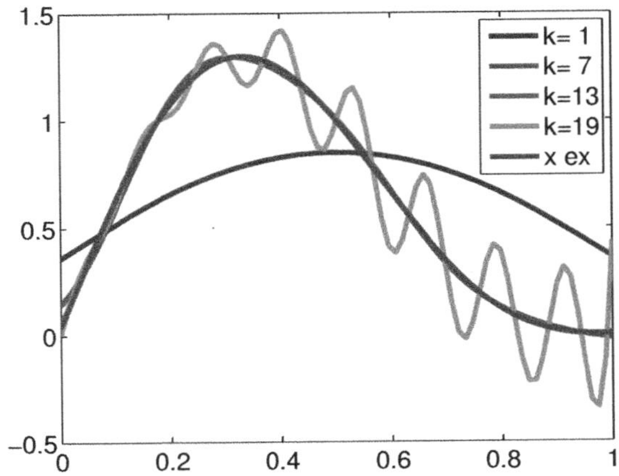

Figure 5.6. *Regularization by spectral truncation for the gravimetric prospecting operator. For a color version of the figure, see www.iste.co.uk/kern/inverse.zip*

5.3. Choice of the regularization parameter

We have discussed in a qualitative manner the influence of the regularization parameter on the quality of the regularized solution. The two pitfalls to avoid are to choose a parameter that is too small (instability dominates) or too large (loss of precision in the solution). The question that then naturally rises is: is it possible to use the data to guide this choice and to obtain a parameter within an acceptable range?

This question remains difficult to answer. Several methods of choice have been proposed and from the point of view of the author, none are truly satisfactory. It remains necessary, in each particular case, to explore a wide range of parameters so that a region that results in a satisfactory solution can be found. Nevertheless, we will present two methods that are quite often used and have the merit of being simple. It should be noted that there exists a method based on the statistical interpretation of the least squares, called *generalized cross-validation* (see [WAH 77, HAN 98]).

5.3.1. *Morozov's discrepancy principle*

We will give here some indications concerning the most conventional method for the choice of the regularization parameter, Morozov's *discrepancy principle* [MOR 84], according to Kirsch [KIR 96].

It is as always assumed that the data available are noisy, and in addition we assume that we have an estimate of the noise level δ. Morozov's discrepancy principle proposes to search for the solution x^{ε} that achieves the minimum of the cost function defined in [5.1], by adding the constraint:

$$\left\|Ax_{\varepsilon}^{\delta} - z^{\delta}\right\|_{2} = \delta, \tag{5.25}$$

which provides an equation relating ε to δ. A heuristic justification for this choice is that it is pointless to reduce the error below the noise level.

By replacing z by z^{δ}, we denote x_{ε}^{δ} the solution of equation [5.3]. We are searching for $\varepsilon = \varepsilon(\delta)$ such that relation [5.25] be satisfied.

PROPOSITION 5.5.– Under the hypothesis $\left\|z^{\delta} - z\right\|_{F} \leq \delta < \left\|z^{\delta}\right\|_{F}$, equation [5.25] admits a unique solution.

Let $x \in E$ such that $Ax = z$, and denote x^{δ} the solution of the corresponding regularized problem, where the regularization parameter ε is chosen as a function of δ by Morozov's discrepancy principle. The method is convergent:

$$\lim_{\delta \to 0} x^{\delta} = x. \tag{5.26}$$

PROOF.– According to proposition 5.3, the function $\varepsilon \to \left\|Ax_{\varepsilon}^{\delta}\right\|_{F}$ is continuous, decreasing with $\lim_{\varepsilon \to \infty} = \left\|z^{\delta}\right\|_{F} > \delta$ and $\lim_{\varepsilon \to 0} = 0 < \delta$. Therefore, there is a single value of ε solution of equation [5.25].

We will prove the second part of the proposition under the additional (regularity) hypothesis $x \in \operatorname{Im} A^*$, with $x = A^* w$. For the general case, see [KIR 96].

According to the definition of x^{δ},

$$\begin{aligned} \left\|Ax^{\delta} - z\right\|_{F}^{2} + \varepsilon(\delta)^{2} \left\|x^{\delta}\right\|_{E}^{2} &\leq \|Ax - z\|_{F}^{2} + \varepsilon(\delta)^{2} \|x\|_{E}^{2} \\ &= \varepsilon(\delta)^{2} \|x\|_{E}^{2} + \left\|z - z^{\delta}\right\|_{F}^{2} \leq \varepsilon(\delta)^{2} \|x\|_{E}^{2} + \delta^{2} \end{aligned}$$

and since $\left\|Ax^{\delta} - z\right\|_{F} = \delta$, we obtain $\left\|x^{\delta}\right\|_{E} \leq \|x\|_{E}$, for any $\delta > 0$. It follows that

$$\left\|x^{\delta} - x\right\|_{E}^{2} = \left\|x^{\delta}\right\|_{E}^{2} - 2(x^{\delta}, x) + \|x\|_{E}^{2} \leq 2\left(\left\|x^{\delta}\right\|_{E}^{2} - (x^{\delta}, x)\right) = 2(x - x^{\delta}, x).$$

Then,

$$\begin{aligned} \left\|x^{\delta} - x\right\|_E^2 &\leq 2(x - x^{\delta},\, A^* w) = 2(z - Ax^{\delta},\, w) \\ &\leq 2(z - z^{\delta},\, w)(z^{\delta} - Ax^{\delta},\, w) \leq 2\delta \left\|w\right\|_F + 2\delta \left\|w\right\|_F, \end{aligned}$$

that is $\left\|x^{\delta} - x\right\|_E^2 \leq \sqrt{2\delta \left\|w\right\|_F}$. □

The discrepancy principle therefore leads to a convergent method. On the other hand, the order of convergence achieved in the previous proof is only $O(\sqrt{\delta})$. We have therefore lost a half-order of the convergence rate and as a result the previous method is not optimal.

For a given value of δ, the determination of ε is equivalent to solving equation [5.25], and we have seen that the function under consideration is monotonic. A first idea is to use Newton's method, but in fact it is preferable to use the following alternative formulation due to Hebden:

$$h(\varepsilon) = \frac{1}{\left\|Ax^{\varepsilon} - \hat{z}\right\|^2} - \frac{1}{\delta^2}.$$

It can be verified that the derivative of the application $\varepsilon \to x_{\varepsilon}^{\delta}$ is obtained by solving the equation:

$$A^* Ax + \varepsilon^2 x = -\varepsilon x_{\varepsilon}^{\delta},$$

and it is then easy to deduce the derivative of h.

The determination of the value of ε satisfying [5.25] therefore requires the solution of a large number of problems such as [5.6], for different values of ε. We outline an effective numerical method in section 5.3.3.

The attentive reader will have noticed that we have replaced the question of the choice of a parameter (the regularization level ε) by another (the noise level δ). He or she may legitimately ask himself (or herself) where the gain is. The answer is that it is probably easier to *estimate* the noise level contained in data (possibly by knowing the experimental device) than to *guess* the regularization value that will lead to a suitable solution. However, the following example shows that this choice has nevertheless an important influence on the result.

EXAMPLE 5.4 (Gravimetric prospecting).– We illustrate the behavior of Morozov's discrepancy method in the geomagnetic prospection example. White noise of variance 10^{-2} can be added to the exact solution, the norm of the perturbation is therefore $\|\hat{z} - z^\delta\|_2 = 10^{-2}$. We have here also used the *Regularization Toolbox*, namely the function `discrep`.

We have initialized the method with the correct value, that is $\delta = 10^{-2}$, (Figure 5.7, on the left) which provides a regularization parameter $\varepsilon \approx 4\,10^{-2}$, and leads to a visually suitable solution. We have then underestimated the estimation of the noise, by taking $\delta = 10^{-3}$, which leads to a much too small choice of the regularization parameter $\varepsilon \approx 10^{-16}$, and to a clearly under-regularized solution (see right of the figure).

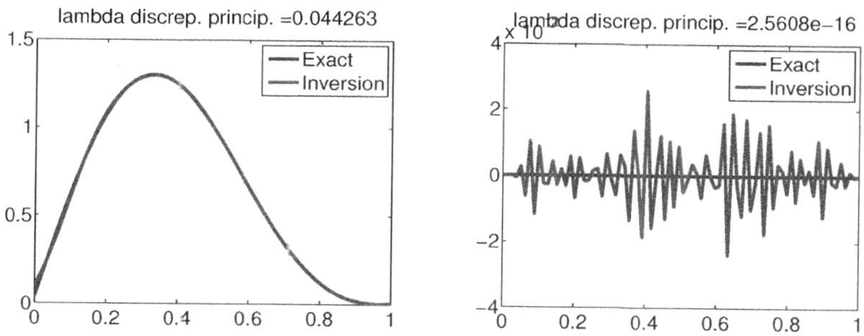

Figure 5.7. *Solutions computed with a parameter chosen by the discrepancy method. On the left $\delta = 10^{-2}$ (exact value); on the right $\delta = 10^{-3}$ (underestimation). For a color version of the figure, see www.iste.co.uk/kern/inverse.zip*

5.3.2. *The L-curve*

The L-curve can also be used as a method to choose the regularization parameter. The intuition is that, if the vertical part of the curve corresponds to too much regularization, and its horizontal part to not enough regulation, the values of the parameters close to the "corner" of the curve should result in sensible solutions. An operational definition of the corner is, according to Hansen [HAN 93b, HAN 98], the point of maximal curvature. These references also propose an algorithm for the computation, based on a smoothing by splines functions of an approximation obtained from a finite number of points, and which is implemented in the function `l_curve` of the *Regularization Toolbox*.

We limit ourselves to illustrating this function in Figure 5.8 (always concerning our working example) in the case of Tikhnov's regularization (left on the figure) and of truncated SVD (on the figure right). We refer the reader to Figures 5.2 and 5.6 to judge the quality of the estimated parameters (in this example, the two solutions are under-regularized).

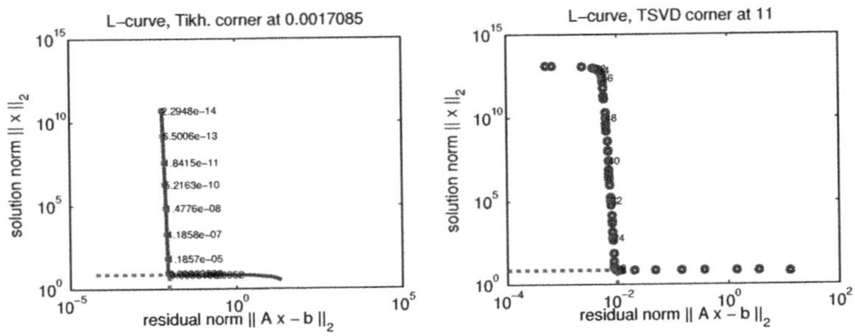

Figure 5.8. *L-curve for the example of geomagnetic prospecting. On the left, Tikhonov's regularization; on the right, truncated SVD. For a color version of the figure, see www.iste.co.uk/kern/inverse.zip*

5.3.3. *Numerical methods*

We give some indications about the numerical methods used to solve the regularized problem [5.1]. The simplest but not the most effective method is to use the form [5.2]. It is equivalent to a standard least squares problem, to which the methods of Appendix 1 can be applied, in particular the QR method.

This approach has the merit of being numerically stable and simple. It has however the disadvantage of requiring the formation of matrix $\tilde{A} = \begin{pmatrix} A \\ \varepsilon I \end{pmatrix}$, of size $(m+n)\times n$. In the case where this computation must be performed repeatedly (to estimate the regularization parameter), the application of this method becomes very expensive. We briefly present an alternative method, whose cost is essentially independent of the number of systems to solve, and which becomes particularly interesting when the objective is to determine the regularization parameter, as in section 5.3.1.

This approach comprises two steps:

– the first step consists of transforming A into a bidiagonal form, in a manner similar to the computation of the singular values of A (see section A1.4);

– we then solve the different problems for each value of the regularization parameter, by a method using Givens rotations (see [GOL 96, DEM 97]).

The first step transforms A into a matrix

$$A = U\begin{pmatrix} B \\ 0 \end{pmatrix} V^T,$$

where $U \in \mathbf{R}^{m\times m}$ and $V \in \mathbf{R}^{n\times n}$ are orthogonal and B is bidiagonal. Problem [5.2] then becomes:

$$\min \left\| \begin{pmatrix} B \\ \epsilon I \end{pmatrix} \xi - \begin{pmatrix} U^T z \\ \varepsilon \xi_0 \end{pmatrix} \right\|_2, \qquad [5.27]$$

where $\xi = V^T x$ and $\xi_0 = V^T x_0$.

Since the matrices U and V are orthogonal, we obtain the relations:

$$\|\xi\|_2 = \|x\|_2, \quad \|Ax - z\|_2 = \left\|B\xi - U^T z\right\|_2.$$

We will not describe the second step in detail here. It should be simply noted that the matrix $\begin{pmatrix} B \\ \epsilon I \end{pmatrix}$ is orthogonally transformed to $\begin{pmatrix} \hat{B} \\ 0 \end{pmatrix}$:

$$\begin{pmatrix} B \\ \epsilon I \end{pmatrix} = Q_\varepsilon \begin{pmatrix} \hat{B} \\ 0 \end{pmatrix}, \qquad [5.28]$$

where $Q_\varepsilon \in \mathbf{R}^{2n\times 2n}$ is orthogonal, and $\hat{B} \in \mathbf{R}^{n\times n}$ is still bidiagonal. Note that the matrix Q_ε is not explicitly formed, but is kept in a factorized form similar to what is done in section A1.3.2 for the matrix Q in the QR decomposition. The initial problem [5.2] is finally equivalent to:

$$\min \left\| \begin{pmatrix} \hat{B} \\ 0 \end{pmatrix} \xi_\varepsilon - \begin{pmatrix} \hat{\beta}_1 \\ \hat{\beta}_2 \end{pmatrix} \right\|_2, \qquad [5.29]$$

whose solution is simply $\xi_\varepsilon = \hat{B}^{-1}\hat{\beta}_1$. Note that the reduction [5.28] and the solution of [5.29] must be carried out for each value of ε, but their complexity is only $O(n)$.

The solution of the initial regularized problem [5.2] is then simply obtained by $x = V\xi_\varepsilon$.

More information on the various methods for the choice of the regularization parameter can be found in Hansen's works. The book [HAN 10] contains a comparison of the various estimation methods, and some practical conclusion, whereas [HAN 98] presents the numerical aspects in more detail.

5.4. Iterative methods

This section is an introduction to a different class of methods for the regularization of linear inverse problems: iterative methods. The methods (Tikhonov's regularization, spectral truncation) we have previously seen are referred to as *direct* because they give (in the case of finite dimension) the *exact* solution of a regularized problem (up to the rounded errors) in a finite number of operations. For moderate sized problems, these methods are the most commonly used. However, for large-scale problems, where the matrices obtained after discretization are often sparse, these methods prove to be inadequate because, on the one hand, the number of operations becomes too large, and on the other hand, these methods do not respect the structure of the matrices.

An alternative is then to resort to iterative methods, which build a *sequence* of approximate solutions that (in the noiseless case) converge toward the desired solution. We will see that in the context of inverse problems, the situation is more complicated: in the presence of noise, the sequence constructed by the iterative method does not converge, in general, toward a solution of the initial problem. Once again, it is necessary to regularize the iterative process, and it is the iteration index itself that plays the role of regularization parameter. In other words, iterations should be stopped earlier than a noiseless case would require.

In this section, we examine the simplest of the iterative methods: the Landweber method [LAN 51], whose main advantage is to lend itself to a simple analysis. Unfortunately, it converges too slowly to be usable in practice, all the more so that much more efficient methods do exist. The two most important are Brakhage's "ν" method (see section 6.3 in [ENG 96], and especially the conjugate gradient method and its variants. This latter method is the most commonly used method. In the context of ill-posed problems, an accessible presentation can be found in Kirch's book [KIR 96], while more complete presentations can be found in [ENG 96, HAN 95] (the latter reference is entirely dedicated to the analysis of conjugated gradient-type methods for ill-posed problems).

Given a relaxation parameter ω, Landweber's method for solving the least squares problem [4.3] is defined by the recurrence formula:

$$x_{n+1} = x_n + \omega(A^\star \hat{z} - A^\star A x_n). \qquad [5.30]$$

We use $x_0 = 0$ for simplicity (the general case is discussed in [KIR 96]). In this case, we can immediately see by induction on n that $x_n \in \operatorname{Im} A \subset \operatorname{Ker} A^\perp$. As a consequence, if x_n converges toward an element $x \in E$, there must hold $x \in \operatorname{Ker} A^\perp$ (since this subspace is closed). In addition, by continuity, x satisfies equation [4.3], and thus $x = \hat{x}$ is the minimal norm solution of [4.2]. Therefore, there only remains to find is a condition on ω that ensures the convergence of x_n to $\hat{x}$, and then to show that, in the noisy case, n can be chosen according to the noise level.

To simplify, we confine ourselves to the case where A is a compact operator, and we note $(\sigma_j)_{j\in\mathbf{N}}$ its singular values, and $(u_j)_{j\in\mathbf{N}}$, $(v_j)_{jn\mathbf{N}}$ its singular vectors The following analysis is essentially that of Groetsch [GRO 93].

PROPOSITION 5.6.– Let A be a compact operator. Landweber's method satisfies the following properties:

1) The iterates satisfy

$$x_n = \omega \sum_{j=0}^{n-1} (I - \omega A^\star A)^j A^\star \hat{z}. \qquad [5.31]$$

2) If we choose $0 < \omega < 2/\sigma_1^2$, the sequence of iterates x_n of Landweber's method converges to $\hat{x}$ for $\hat{z} \in \operatorname{Im} A$.

3) Under the additional (regularity) hypothesis $\hat{x} \in \operatorname{Im} A^\star$, we have in addition the estimate

$$\|x_n - \hat{x}\| = O(1/n). \qquad [5.32]$$

PROOF.–

1) Equation [5.31] can be easily proved by induction.

2) Denote by $e_n = x_n - \hat{x}$ the error. By subtracting [5.30] from [4.3], for $n > 0$ we obtain:

$$e_n = (I - \omega A^\star A) e_{n-1} = (I - \omega A^\star A)^n e_0. \qquad [5.33]$$

If we introduce the singular value expansion of $e_0 = \sum_{j=1}^{\infty} (e_0, u_j) u_j$, equation [5.33] can be rewritten:

$$e_n = \sum_{j=1}^{\infty} (e_0, u_j)(1 - \omega\sigma_j^2)^n u_j.$$

Since the singular vectors are orthogonal, we obtain:

$$\|e_n\|^2 = \sum_{j=1}^{\infty} |(e_0, u_j)|^2 \, |1 - \omega\sigma_j^2|^{2n}, \qquad [5.34]$$

and if $\omega < 2/\|A\|^2 = 2/\sigma_1^2$, each term $|1 - \omega\sigma_j^2|$ is in absolute value strictly less than 1. To pass to the limit in [5.34], we use Lebesgue's dominated convergence theorem. First, each term of the sum tends to 0 and is bounded by $|(e_0, u_j)|$. But, according to Bessel's inequality,

$$\sum_{j=1}^{\infty} |(e_0, u_j)|^2 \leq \|e_0\|^2 .$$

Lebesgue's theorem makes it thus possible to conclude that $e_n \to 0$, and therefore that the sequence x_n converges to $\hat{x}$.

3) By hypothesis, there exist $w \in E$, such that $\hat{x} = A^\star w$ and [5.34] becomes:

$$\|e_n\|^2 = \sum_{j=1}^{\infty} \sigma_j^2 \, |(w, v_j)|^2 \, \left|1 - \omega\sigma_j^2\right|^{2n} .$$

Then, using the inequality (proved below)

$$\left|1 - \omega\sigma_j^2\right|^n \leq \frac{1}{n\omega}\frac{1}{\sigma_j^2}, \qquad [5.35]$$

we obtain:

$$\|e_n\|^2 \leq \frac{1}{n^2\omega^2} \sum_{j=1}^{\infty} |(w, v_j)|^2 = \frac{1}{n^2\omega^2} \|w\|_F^2 \qquad [5.36]$$

This is the estimate of the proposition. □

Let us now prove inequality [5.35], as well as another inequality that will be useful for the proof of theorem 5.2.

LEMMA 5.3.– We have the inequalities, valid for any $x \in]0, 1[$

$$(1 - x)^n \leq \frac{1}{nx} \qquad [5.37]$$

$$1 - (1 - x)^n \leq nx \qquad [5.38]$$

PROOF.– For the first inequality, it suffices to show that

$$nx(1-x)^n \leq 1, \quad \text{for } x \in [0,1].$$

The function under consideration is positive, its values at 0 and 1 are both zero and its derivative vanishes at the point $1/(n+1)$, where the function is equal to $\left(\frac{n}{n+1}\right)^{n+1} \leq 1$.

The second inequality is a consequence of the concavity of the function $x \mapsto (1-x)^n$:

$$1-(1-x)^n \leq 1-(1-nx) = nx \qquad \square$$

Therefore, we can see that the iterative method converges for noiseless data. Now consider what happens if we replace z by a noisy datum z^δ, with $\left\|z - z^\delta\right\| = \delta$. We denote $x^{n,\delta}$ the sequence of corresponding iterates and we introduce the sequence x^n constructed from noiseless data z. We want to estimate the error $x^{n,\delta} - \hat{x}$, and show that we can choose n according to δ to obtain convergence.

THEOREM 5.2 (Regularization strategy for Landweber's method).– Under the same hypotheses as proposition 5.6, we have the following:

1) If n is chosen such that $\underset{n\to\infty}{n(\delta) \to 0}$, then $\lim_{\delta\to 0} x^{n(\delta),\delta} = \hat{x}$.

2) Under the additional hypothesis $\hat{x} \in \operatorname{Im} A^\star$, we can choose $n(\delta) = \lceil(1/\sqrt{\delta})\rceil$ and then $\left\|x^{n(\delta),\delta} - \hat{x}\right\| = O(\sqrt{\delta})$ ($\lceil . \rceil$ is the *ceiling* function).

PROOF.– Once again, we use the triangle inequality to write:

$$\left\|x^{n,\delta} - \hat{x}\right\| \leq \|x^n - \hat{x}\| + \left\|x^{n,\delta} - x^n\right\|$$

In the first case, we know that the first term tends to 0; in the second, we know in addition that it behaves (at least) as $O(1/n)$.

To estimate the second term, we subtract the equation defining $x^{n,\delta}$ from that defining x^n. Noting $d_n = x^{n,\delta} - x^n$, we find:

$$d^{n+1} = (I - \omega A^\star A)d^n + \omega A^\star(z^\delta - z), \; d_0 = 0.$$

This recurrence is the same as that which defines the original sequence x_n. We can thus use [5.31]. A similar reasoning to that of proposition 5.6 leads to the representation:

$$\left\|d^{n+1}\right\|^2 = \sum_{j=1}^{\infty} \frac{1-(1-\omega\sigma_j^2)^{2n}}{\sigma_j^2} \left|(z^\delta - z, v_j)\right|^2. \qquad [5.39]$$

that is, according to lemma 5.3:

$$\left\|d^{n+1}\right\|^2 \leq \omega n \sum_{j=1}^{\infty} \left|(z^\delta - z, v_j)\right|^2 = n\omega \left\|z^\delta - z\right\|^2 \leq n\omega\delta. \quad [5.40]$$

1) In the general case, we only have:

$$\left\|x^{n,\delta} - \hat{x}\right\| \leq \|x^n - \hat{x}\| + n\omega\delta$$

and the first term tends toward 0. The condition $n\delta \to 0$ is sufficient to ensure convergence.

2) In this case, the rate of convergence can be made precise using (ii) of the previous theorem.

$$\left\|x^{n,\delta} - \hat{x}\right\| \leq \frac{1}{n\omega} \|w\|_F^2 + n\omega\delta.$$

The sum on the right-hand side is minimal when $n\omega = \|w\|_F \delta^{-1/2}$, and the minimum is equal to $\|w\|_F \delta^{1/2}$. □

As we had mentioned at the beginning of this section, in order to obtain the convergence of the iterative method, we need to stop the iterations at a level dependent on the noise level: the sooner the better as the noise becomes stronger. Moreover, for Tikhonov's method, this result assumes that we know how to estimate the noise level. Finally, we will once more draw the attention of the reader to the fact that Landweber's method converges too slowly to be usable in practice. There are other iterative methods, such as the conjugate gradient method, that converge much faster for an essentially equivalent cost.

5.5. Exercises

EXERCISE 5.1.– Regularization of the derivative.

Tikhonov's method is applied to the integral equation:

$$\int_0^t x(s)\,ds = z(t), \quad 0 \leq t \leq 1.$$

Show that, for $z \in H^1(0,1)$, with $z(0) = 0$, the regularized solution u_ε is the solution of the boundary problem:

$$-\varepsilon^2 x''(t) + x(t) = z'(t), \quad 0 < t < 1, \quad x(1) = 0,\ x'(0) = 0.$$

EXERCISE 5.2.– Regularization by filtering.

Let E and F be two Hilbert spaces,$(e_j)_{j\in\mathbf{N}}$ and $(f_j)_{j\in\mathbf{N}}$ be Hilbert bases of E and F, respectively. Let $A \in \mathcal{L}(E, F)$ be the compact operator defined by:

$$Ae_j = \sigma_j f_j, \quad A^* f_j = \sigma_j e_j, \qquad \forall j \in \mathbf{N}.$$

where $(\sigma_j)_{j\ in\mathbf{N}}$ is the sequence of singular values ($\sigma_j > 0$ and $\lim_{j\to\infty} \sigma_j = 0$) of A.

Let $x \in E$ and $z \in F$ be such that $Ax = z$, and let z^δ such that $\left\|z^\delta - z\right\|_F = \delta > 0$. We thus have:

$$x = \sum_{j\in\mathbf{N}} x_j e_j, \; z = \sum_{j\in\mathbf{N}} z_j f_j, \text{ with } z_j = \sigma_j x_j.$$

To regularize the problem $Ax = z$, we consider the regularization method:

$$x^{\varepsilon,\delta} = \sum_{j\in\mathbf{N}} q_\varepsilon(\sigma_j)\frac{1}{\sigma_j} z_j^\delta e_j,$$

where $q : \mathbf{R}^+ \to \mathbf{R}$ is a function that satisfies the following hypotheses:

i) $|q_\varepsilon(\sigma)| \leq 1, \quad \forall\varepsilon > 0, \forall\sigma > 0.$

ii) There is a function $c(\varepsilon)$ such that $|q_\varepsilon(\sigma)| \leq c(\varepsilon)\sigma, \quad \forall\sigma > 0.$

iii) There exists $c_1 > 0$ such that $|q_\varepsilon(\sigma) - 1| \leq c_1\frac{\varepsilon}{\sigma}, \quad \forall\varepsilon > 0, \forall\sigma > 0.$

1) Show that Tikhonov's method and the truncated SVD methods are special cases of the previous method.

2) To estimate the error $\left\|x^{\varepsilon,\delta} - x\right\|_E$, we introduce $x^\epsilon = \sum_{j\in\mathbf{N}} q_\varepsilon(\sigma_j)\frac{1}{\sigma_j} z_j e_j$ the regularized version of the exact solution. Estimate separately each of the differences $\left\|x^{\varepsilon,\delta} - x^\varepsilon\right\|_E$ and $\|x^\varepsilon - x\|_E$.

3) Find again the regularization strategies (*a priori*) for the Tikhonov's and the spectral truncation methods, that is a function $\delta \to \varepsilon(\delta)$.

4) Now consider the finite dimension case ($E = \mathbf{R}^n, F = \mathbf{R}^m$). Using the SVD of A, show that the the solution of the regularized problem can be written in the form:

$$x = V\,(\Phi_\varepsilon\;0)\,U^T z,$$

where Φ_ε is a matrix that will be calculated.

5) Consider Landweber's iterative method:

$$x^{k+1} = (I - \omega A^* A)x^k + \omega A^* z, \quad x^0 = 0.$$

Calculate the associated function q. How is the regularization achieved?

EXERCISE 5.3.– Regularization and singular values for an integral Volterra operator.

Consider the integral operator A on $L^2(0,1)$ defined by:

$$Au(t) = \int_0^t (t-s)u(s)\,ds, \quad t \in]0,1[, \qquad [5.41]$$

and the associated integral equation:

$$Au = f, \quad f \text{ given in } L^2(0,1). \qquad [5.42]$$

Recall that Sobolev's space $H^2(0,1)$ is defined by:

$$H^2(0,1) = \{u \in L^2(0,1),\, u' \in L^2(0,1),\, u'' \in L^2(0,1)\}$$

1) Show that equation [5.42] is equivalent to the Cauchy problem for the differential equation:

$$f''(t) = u(t),\; t \in]0,1[, \quad f(0) = f'(0) = 0. \qquad [5.43]$$

Determine then the kernel and the image of A.

2) Compute the adjoint of A. Show that equation $A^*v = g$ is equivalent to the following problem:

$$g''(t) = v(t),\; t \in]0,1[, \quad g(1) = g'(1) = 0. \qquad [5.44]$$

3) It is assumed here that $f \in \mathrm{Im}A$. Equation [5.42] is regularized using Tikhonov's method. By decomposing the regularized normal equation $A^*Au + \varepsilon^2 u = A^*f$ into a coupled system (r is the residual):

$$\begin{cases} Au + r = f \\ -\varepsilon^2 u + A^* r = 0 \end{cases} \qquad [5.45]$$

reduce the normal equation to a boundary value problem for a fourth-order equation. Specify the boundary conditions. The solution of this boundary problem is *not* required.

4) Using a similar method, write the boundary problem satisfied by a singular triplet (σ, u, v):

$$Au = \sigma v, \qquad A^* v = \sigma u \qquad [5.46]$$

The solution of this equation is not required.

EXERCISE 5.4.– Introduction to the *Regularization Toolbox* software.

This (long) exercise aims to illustrate the use of the theoretical concepts met along this chapter on a simple example using the *Regularization Toolbox*. This software, written in MATLAB, is written by Hansen. It is available on the Website of the author at the URL: http://www2.imm.dtu.dk/pcha/Regutools/. You will have to extract the files from the archive to a local folder and add the corresponding path to the *path* of MATLAB.

A manual is included in the distribution and online help is available in MATLAB. You can see a demo of the main functions of the software by issuing the command: `regudemo`.

For the following, we choose the test problem "Baart". This is an integral equation:

$$\int_0^{\pi/2} \exp(s\cos(t))\, u(t)\, dt = 2\sinh(s)/s \quad t \in [0, \pi]$$

discretized by a Galerkin method, using functions with discontinuous bases. The exact solution is $u(t) = \sin t$.

1) Compute the matrix and the right-hand side of the system. The corresponding MATLAB command is (we can choose $n = 40$): `[A,bex,xex] = baart(n)`.

2) Try to solve the linear system by the usual methods (`xnaive= A \bex`). Comment on the obtained solution.

3) Compute the singular value decomposition of A using the command `[U, s, V]= csvd(A)`, then visualize the decrease of the singular values (command `picard`). A few singular vectors can also be visualized. What can be observed?

Add random Gaussian noise of mean 0 and variance η to the right-hand side (set, e.g., $\eta = 10^{-3}$). The corresponding command is (relative perturbation): `e = eta*randn(size(bex)); b = bex .* (1+ e).`

Re-examine to the previous question.

4) Choose a few values for the regularization parameter ε and compute the corresponding regularized solution (command `tikhonov`; note that for this command, the regularization parameter may be a *vector*, which makes it possible to address all parameters at once).

Do the same for the spectral truncation method (command `tsvd`), by varying the number of retained singular values.

5) Evaluate, for this problem, various methods for the choice of the regularization parameter (commands `l_curve`, `discrep` or `gcv`).

In the case of the discrepancy principle, take $\sqrt{n}\eta$ with $\eta = 10^{-3}$ as an estimation of the noise norm (for the example above), then try increasing or decreasing the value of η by 5%. Evaluate the sensitivity of the solution to these changes.

For the L-curve, or the GCV, generate several different perturbations and draw the corresponding curves. Is the minimum of the functions always found at the same place ("corner" of the L-curve, transition region for GCV)?

PART 3

Nonlinear Inverse Problems

6

Nonlinear Inverse Problems – Generalities

We now address nonlinear inverse problems and focus on parameter identification in differential equations or partial differential equations.

In the general case, we will be confronted with a physical phenomenon whose *structure* is known but whose precise operational parameters are not. It is possible to measure certain properties of this system, corresponding to known entries. The system works like a black box, and we want to know what it contains without "opening" it.

We have seen examples in Chapter 2. We recall example 1.2: we *know* (or, more accurately, we assume) that heat conduction obeys equation [2.7], but the coefficient K is not known (it can be a scalar or a function of position). On the other hand, we assume that we have access to a *measurement* of the temperature T, in part of the domain (or the boundary). With this information, we want to determine the coefficient K that can reproduce these measurements.

The first difference with the previous chapters is that the mapping from the parameter (c in our example) to the measurement is not only nonlinear, but is expressed by means of an equation such as [2.7] and the measurement is then a "part" of the solution of this equation. An equation that plays the same role as [2.7] is called a *state equation* and the variable T, a solution of this equation, is called the *state* of the system. In general, it will be unrealistic to assume that the whole state of the system is known: still with our thermal example in mind, we see that it is not possible to measure the temperature at all points of the domain.

A second, more practical, difference lies in the fact that it is more difficult to obtain theoretical results than in the linear case. The results that have been obtained are often linked to a particular problem. We refer to the books [CHA 09] and [BAN 89] for elements of a general theory, as well as Chavent's work (e.g. [CHA 91]). In any case, in the context of this book we will not discuss these issues (which does not mean that they are less important), and we will focus on the numerical methods, and in particular on the least squares formulation.

Chavent's book is, to our knowledge, the only book that addresses in detail the computational methods for nonlinear inverse problems; in particular, the problem of computing the gradient by the adjoint state technique (to which we dedicate section 6.3.3 and Chapter 7). The present book can be seen as a preliminary to the reading of [CHA 09], which presents a more general, but more abstract, vision of the subject.

6.1. The three fundamental spaces

In order to provide an abstract formulation of the problems that we will consider later in this book, we introduce three Hilbert spaces, as well as some mappings between these spaces. In the examples that follow, we will always end up performing the computations in finite-dimensional spaces. Nevertheless, these spaces will be approximations of "infinite-dimensional" spaces, and it will be important not to make use of the particular properties of these approximations. It thus seems appropriate to provide the definitions within this more general context. The three spaces are as follows:

– the *model* (or parameter) space M;

– the *state* space U;

– the *data* (or observations) space D.

As we have pointed out, the introduction of the space U makes the dependency between the parameter and the data explicit. On the other hand, the existence of the state space does not exempt us from introducing the observation space, since in general the state cannot be directly measured.

Two mappings emphasize the relationships between these three spaces:

– the state equation implicitly relates the parameter and the state (both can obviously be vectors). We write it as:

$$F(a,u) = 0, \quad a \in M,\ u \in U,\ F(a,u) \in Z, \qquad [6.1]$$

where Z is another Hilbert space. We assume that there is a subspace $M_{\text{ad}} \subset M$, the subset of the admissible parameters, such that for any $a \in M_{\text{ad}}$, F locally defines a unique state $u = u_a$.

It will be convenient to denote

$$u = S(a) = u_a \qquad [6.2]$$

the solution of the state equation [6.1]. The first equality will be helpful when we have to operate on this solution (to differentiate it, for example), while the second is a suggestive abuse of notation;

– the observation operator extracts from the state the part corresponding to the measurements. This operator will often be an injection, rarely the identity (except in purely pedagogical examples). It is written as:

$$d = Hu, \qquad u \in U. \qquad [6.3]$$

We made the simplifying hypothesis that the observation is a linear operator, independent of the parameter. Extending these considerations to the more general situation is not difficult and this is left to the reader.

If we insert the solution of [6.1] in [6.3], we obtain the mapping that relates the parameter to the observation. We write this mapping as:

$$d = \Phi(a) = H\big(S(a)\big) = H(u_a). \qquad [6.4]$$

The inverse problem is then:

given an observation d_{obs}, solve the equation $\Phi(a) = d_{\text{obs}}$. [6.5]

We give a few examples to illustrate the previous concepts. We will return in more detail to these examples later in this chapter and in the following chapter, to gradients the corresponding functionals, and then calculate their gradients.

We start with a simple example, without any real physical meaning.

EXAMPLE 6.1.– We consider the boundary value problem in one space dimension:

$$\begin{cases} -bu''(x) + cu'(x) = f(x) \quad 0 < x < 1, \\ u(0) = 0, \ u'(1) = 0. \end{cases} \qquad [6.6]$$

In this case, the parameter is the pair $a = (b, c)$ and $M = \mathbf{R}^2$. The Lax–Milgram theorem shows that a natural choice for U is $U = \big\{u \in H^1(0,1), u(0) = 0\big\}$ and that we can choose $M_{\text{ad}} = \{(b,c) \in M, \, b > 0\}$. We consider several possibilities for the observation:

1) it is first assumed that the state u is measured at all points of the interval $]0,1[$. The space D is then $L^2(0,1)$ and d_{obs} is a function defined in $]0,1[$. In this case, this inverse problem is overdetermined. We have here $Hu = u$;

2) at the other extreme, it can be assumed that u is only measured at the right end of the interval. In this case, $D = R$, and d_{obs} is a number. This problem is here *under-determined*. This time, we have $Hu = u(1)$;

3) an intermediate case is one where u is measured not only at the end of the interval, but also at one interior point. To fix ideas, we will assume that u is known at $x = 1/2$. The space D is here $D = \mathbf{R}^2$, and d_{obs} is a pair, and $Hu = (u(1/2), u(1))$. In this case, there is exactly the same amount of data as of unknowns. This does not make the problem any simpler, since the ill-posed nature of the inverse problem must always be taken into account.

In all three cases, the mapping Φ is defined by (we should recall that $u_{(b,c)}$ denotes the solution of the state equation [6.6] corresponding to the parameters (b,c)):

1) $\Phi(b,c) = u_{(b,c)}$ (function defined in $]0,1[$);

2) $\Phi(b,c) = u_{(b,c)}(1)$;

3) $\Phi(b,c) = (u_{(b,c)}(1/2), u_{(b,c)}(1))$.

In the latter two cases, the mapping Φ goes from one finite dimensional space to another, but its definition involves the solution of the boundary problem [6.6].

EXAMPLE 6.2.– We now consider a time-varying system, which we will assume as governed by a system of ordinary differential equations:

$$\begin{aligned} y'(t) &= f(y(t), a) \quad t \in [0,T] \\ y(0) &= y_0, \end{aligned} \qquad [6.7]$$

where y takes its values in $\mathbf{R}^d$, the parameter a is in $\mathbf{R}^m$ and $f \in C^1(\mathbf{R}^d, \mathbf{R}^p)$. We will assume that y is measured at certain instants $\tau_1, \ldots, \tau_Q$. The inverse problem is to identify a.

We thus have $M = \mathbf{R}^m$. According to the Cauchy–Lipschitz theorem [ARN 12], the Cauchy problem [6.7] has a unique solution in $C^1(0,T)^d$. This space is not a Hilbert space, but we can take $U = H^1(0,T)^d$.

The observation space is $D = \mathbf{R}^Q$, and the observation operator is:

$$H : y \mapsto (y(\tau_1, a), \ldots, y(\tau_Q, a)).$$

The observation will be a Q-tuple of vectors of $\mathbf{R}^d$, each element representing a measurement instant.

In this example, the mapping Φ again goes from a finite-dimensional space to another (in this case, from $\mathbf{R}^n$ to $\mathbf{R}^Q$), but its definition involves, as in the previous example, solving the differential equation [6.7].

The following example is a generalization of example 6.1 and this time corresponds to a realistic application.

EXAMPLE 6.3 (Elliptic equation, continuation of example 2.2).– In this case, the parameter is the conductivity K. This time, the parameter does not "live" in a finite-dimensional space any longer, but is a function of the spatial position. Obviously, after discretization, we will have a finite number of parameters, but the actual number precisely depends on the fineness of the discretization, which is supposed to tend toward 0 (and the number of parameters tends to infinity). It is important to take into account the character inherently *distributed* of the parameter to be identified and to postpone the choice of a finite-dimensional representation as late as possible.

To effectively show that this example falls within the general context at the beginning of this section, we will rewrite the equations of example 2.2 with abstract "notations". We will therefore denote by a the conductivity, and by u the temperature. The state equation is thus:

$$\begin{cases} -\operatorname{div}(a \operatorname{grad} u) = f(x, y, z) & \text{in } \Omega \\ u = 0 & \text{on } \partial\Omega \end{cases} . \tag{6.8}$$

A possible choice for the space M is $L^2(\Omega)$, and

$$M_{\text{ad}} = \left\{ a \in L^2(\Omega), \exists (a_m, a_M),\ 0 < a_m \leq a(x) \leq a_M < \infty \,(\text{pp}) \right\},$$

which is a convex subset (but not a vector subspace) of M. The space U is naturally $H^1(\Omega)$, so that for $a \in M_{\text{ad}}$, the existence of $u = u_a$ is ensured by the Lax–Milgram theorem [DAU 90, vol. 4].

In order to specify the observation operator, we will consider several situations:

1) The simplest case (but as we have already pointed it out, it is unrealistic) is when the state u is measured everywhere. Even in this case, the operator H will not be the identity: because of the choice of U, it would mean that we measure u *and its derivative*. We will therefore assume that D is the space $L^2(\Omega)$, and that H is the canonical injection from U to D. The observation is then a function $d_{\text{obs}} \in L^2(\Omega)$, that is a function defined in Ω.

2) A more realistic case is that in which the temperature is measured only on the boundary of the domain. We take in this case $D = L^2(\Gamma)$ and the observation operator is the trace operator on the boundary (it is well defined on U). The observation this time is a function defined on $\partial\Omega$.

3) Finally, for a third example we assume that we measure the temperature at fixed points $x_q \in \Omega, q = 1, \ldots, Q$. Such an observation is not well defined in our mathematical context, since the functions in U need not be continuous. Moreover, such a measurement is never actually pointwise in practice, it is always an average of the temperature around the point being considered. Let thus ω_q be a neighborhood of the point x_q, we take $\int_{\omega_q} u(x)\, dx$ as a measurement, which is well defined if ω_q is bounded. The space D is then $\mathbf{R}^Q$, and the observation is the mapping

$$u \mapsto \left(\int_{\omega_q} u(x)\, dx \right)_{q=1,\ldots,Q}.$$

In this case, the observation d_{obs} is a vector.

In the first two cases, the observation is *continuous*. It is distributed in the first case, and of boundary type in the second case. Finally, in the third case, the observation is *discrete*.

The last example combines characteristics from the previous two: this is an evolutionary partial differential equation, in which the parameter to be identified is a function.

EXAMPLE 6.4 (Heterogeneous heat equation).– The heat equation is the basic model that governs diffusion phenomena and is encoutered in a number of areas in physics (refer back to section 2.1). Given a domain $\Omega \subset \mathbf{R}^2$ (to fix ideas), whose boundary is denoted by Γ, and a real number $T > 0$, we consider the problem:

$$\begin{cases} \dfrac{\partial u}{\partial t} - \operatorname{div}(a \operatorname{grad} u) = f & \text{in } \Omega \times]0, T[, \\ u(x, t) = 0 & \text{in } \Gamma_D \times]0, T[, \\ a \dfrac{\partial u}{\partial n} = g & \text{on } \Gamma_N \times]0, T[, \\ u(x, 0) = u_0(x) & \text{in } \Omega, \end{cases} \qquad [6.9]$$

where $f \in L^2(0, T;\ L^2(\Omega))$, $g \in L^2(0, T;\ L^2(\Gamma_N))$, and $u_0 \in L^2(\Omega)$ are given and assumed to be known functions (Γ_N and Γ_D form a partition of Γ), and we want to identify the function a. With the now usual reservations about its non-Hilbertian nature, the natural choice for the space M is $M = L^\infty(\Omega)$, and $M_{\text{ad}} = \left\{ a \in M,\ a(x) \geq a_* > 0 \right\}$.

We make the hypothesis that u is measured over the part Γ_N of the boundary (and that this observation is continuous in time), and also that $u(x, T)$ is known over all Ω at the final instant. Under these conditions, the data consist of two functions: $\hat{d}_N \in L^2(0, T;\ \Gamma_N)$ and $\hat{d}_T \in L^2(\Omega)$, and the space D is $L^2(0, T;\ \Gamma_N) \times L^2(\Omega)$.

6.2. Least squares formulation

In the examples that we have just discussed, the application Φ is implicitly defined. It is nonlinear, even if the state equation and the observation operator are linear. This obviously makes the solution of the inverse problem more difficult. On the other hand, in examples 2.12 and 2.9, the mapping Φ was explicitly defined, without resorting to an internal state and it was linear.

The preceding considerations suggest that equation [6.4] may not have a solution, and that even if it has any, the inverse mapping is not necessarily continuous. We will therefore introduce an *a priori* weaker formulation that has demonstrated its practical value. We *replace* equation [6.4] by the following *minimization* problem,

$$\text{minimize } J(a) = \frac{1}{2}\left\|\Phi(a) - d_{\text{obs}}\right\|_D^2 \text{ for } a \in M_{\text{ad}}. \qquad [6.10]$$

This formulation is called an *output least squares* method, and J is the *cost function*, or *error functional*. It is important to understand how this reformulation "works". An *experimental measurement* being given once for all, in order to evaluate the functional J at a parameter a, we start by solving the state equation [6.1], and then the observation operator [6.3] is evaluated, and the simulated measurement is compared to the experimental measurement.

REMARK 6.1.– The equation error.

There is another way to formulate an inverse problem as an optimization problem. It consists of replacing the state by the observation (at the risk of interpolating this latter). This leads to a functional

$$J_{\text{eqn}}(a, d) = \frac{1}{2}\left\|F(a, \tilde{d})\right\|^2, \qquad [6.11]$$

where $\tilde{d}$ is a interpolate of d. Since the functional J_{eqn} is quadratic with respect to the parameters, this method is very popular with physicists and engineers. Its main disadvantage is that it needs to interpolate the observation.

We will now go back to the examples of section 6.1 and propose for each one of them a formulation in terms of functional minimization.

EXAMPLE 6.5 (Continuation of example 6.1).– We examine the different observations considered when we introduced this example and propose a functional in each case:

1) When the state is measured over the entire interval, $D = L^2(0,1)$ and it is natural to choose:

$$J_1(b,c) = \frac{1}{2}\int_0^1 |u(x) - d_{\text{obs}}(x)|^2\, dx. \qquad [6.12]$$

2) If we measure u only at the right end of the interval, we will choose:

$$J_2(b,c) = \frac{1}{2}|u(1) - d_{\text{obs}}(1)|^2. \qquad [6.13]$$

3) Finally, if u is measured at two points, we will choose:

$$J_3(b,c) = \frac{1}{2}\left(|u(1/2) - d_{\text{obs}}(1/2)|^2 + |u(1) - d_{\text{obs}}(1)|^2\right). \qquad [6.14]$$

EXAMPLE 6.6 (Continuation of example 6.2).– As a cost function, we consider:

$$J(a) = \frac{1}{2}\sum_{q=1}^{Q} \|y_a(\tau_q) - d^q_{\text{obs}}\|^2_{\mathbf{R}^d}\, \Delta t_q, \qquad [6.15]$$

where $\Delta t_q = \tau_{q+1} - \tau_q$ is the interval between two observation instants. Its introduction makes the sum similar to an integral.

EXAMPLE 6.7 (Continuation of example 6.3).– We propose a cost function for each of the three situations met in example 6.3:

1) In the case where u is measured everywhere (the observation operator is the canonical injection from $H^1(\Omega)$ to $L^2(\Omega)$), we define:

$$J_1(a) = \frac{1}{2}\int_\Omega |u_a(x) - d_{\text{obs}}(x)|^2\, dx. \qquad [6.16]$$

2) In the case where u is measured on the boundary, the natural choice is:

$$J_2(a) = \frac{1}{2}\int_{\partial\Omega} \left|u_{a|\partial\Omega}(x) - d_{\text{obs}}(x)\right|^2\, d\gamma(x). \qquad [6.17]$$

3) Finally, if u is measured in the neighborhood of points of Ω, we take:

$$J_3(x) = \frac{1}{2}\sum_{q=1}^{\mathcal{C}} \int_{\omega_q} \left|u_a(x) - d^q_{\text{obs}}(x)\right|^2 dx. \qquad [6.18]$$

EXAMPLE 6.8 (Continuation of example 6.4).– As in the previous cases, we aggregate the measurement errors in a single cost function:

$$J(a) = \frac{1}{2}\int_0^T \int_{\Gamma_N} \left|u_a - \hat{d}_N\right|^2 dx\, dt + \frac{1}{2}\int_\Omega \left|u_a(x,T) - \hat{d}_T\right|^2 dx. \qquad [6.19]$$

What does such a reformulation bring? Clearly, it cannot change an ill-posed problem into a well-posed problem as if by magic. On the other hand, often makes it possible to restore the existence of a solution. Even if there is no solution to equation [6.4], it will be sometimes possible to show that the minimization problem admits a solution. The essential difficulty remains in showing that the minimum is reached. There are also counter-examples that can be found in [BAN 89]. As a matter of fact, even if the admissible subset is convex, its image by the nonlinear mapping Φ will generally not be convex. The minimization problem [6.10] is interpreted as the projection on a non-necessarily convex set. The *quasi-convexity* theory, exposed in the already cited book by Chavent [CHA 09], allows these problems to be addressed at the theoretical level. Another essential question is that of uniqueness. It can be easily seen that this is linked to the convexity of the functional J. Yet again, this property is not guaranteed.

The formulation [6.10] has, however, some advantages:

– it provides a systematic way for formulating inverse problems;

– in some cases, some properties of the functional J can be demonstrated;

– this formulation makes it also possible to regularize the problem, that is to approach it by a family of well-posed problems, whose solution converges to the solution of the original problem. Elements can be found about the regularization of nonlinear problems in [CHA 09] and [ENG 96];

– there are robust numerical methods that are effectively studied to solve optimization problems (see reminders in Appendix 2);

– under reasonable assumptions about the data, the functional J is differentiable and lends itself to an approach by a local optimization gradient-type method. The remainder of this book is dedicated to this topic.

REMARK 6.2.– On the choice of norms.

We have been working from the beginning with Hilbert spaces and therefore Hilbertian norms. In practice, this is reflected in function spaces such as L^2 or Sobolev spaces. Other norms than a Hilbert space norms could be chosen, and the present choice is essentially made for convenience. In a certain sense, it is not natural, since the parameters often vary in (a subspace of) L^∞. The main advantage of Hilbertian norms is that they lead to situations where we know how to perform the computations. The choice of L^2 norms can also be justified by statistical considerations that we will not address.

On the other hand, a Hilbertian norm has the disadvantage of giving more weight to outlier points, which would not be the case for a norm of the L^1 type. This has led to inversion methods known as "robust" (see Chapter 8, and exercise 6.1).

6.2.1. *Difficulties of inverse problems*

The difficulty of inverse problems comes from a combination of factors:

– as we mentioned in section 6.2, the cost function is generally *non-convex*. This leads to the existence of local minima and the optimization method can converge to any of these minima;

– the inverse problem may be *under-determined*, due to lack of data (which is intrinsic to the problem). This leads to the existence of several solutions, in other words to several parameters producing the same observations;

– the lack of continuity produces *instability*. Even if it is possible (in theory) to solve the problem with exact observations, this does not mean that we can solve it with noisy data, even if the noise level is low;

– a challenge of a different nature is related to the cost of the solution, assuming that it is possible to overcome the previous obstacles. In effect, the simple *evaluation* of the cost function requires the solution of the state equation, that is in general of one (or several) partial differential equation.

6.2.2. *Optimization, parametrization, discretization*

The optimization problem [6.10] as well as the state equation [6.1] are generally defined "in infinite dimension", that is to say that the Hilbert spaces M, U and D are of infinite dimension. They are generally function spaces, as we have seen in the examples of Chapter 2. Before starting to compute a solution on a computer, it is obviously necessary to reduce the problem to a finite dimensional one. Since there is usually only a finite number of observations available, the space D is often of finite dimension from the beginning.

It will naturally be important to discretize the parameter space and the state space (as well as the state equation). It is possible, and often useful, to perform the study of the inverse problem in a functional context (for this approach, see [CHA 09]). It is the only way to obtain theoretical results about the initial problem and this also provides a guide on the discrete computation. However, in general, this requires sophisticated mathematical tools. As we do not wish to cover these tools in detail, we will perform all of the computations after discretization. In addition, there is another reason for proceeding in this manner: this allows to minimize exactly the discrete cost function, which is preferable to the discretization of the continuous cost function.

We will start by discretizing the state equation. As a result, U is replaced by a finite dimension space. This choice obviously has a crucial influence, because changing it implies having to restart the entire analysis process. An optimization problem is thus obtained for which the unknown is still infinite dimensional, but the state equation is defined in finite dimension. This makes it possible to obtain the exact gradient of the cost function actually used by the program. An alternative would be to discretize the state equation at the last moment and to also discretize the continuous gradient. As indicated above, experience shows that this second approach degrades the convergence of the optimization methods and we will continue by computing the exact gradient of the aaproximated functional.

The discretization of the state equation in general imposes a representation of the parameter compatible with this discretization. But it may be desirable to use a different representation of the parameter. The process that consists of replacing M by a finite-dimensional space is known as *parametrization*. Commonly used examples are piecewise constant functions (a cellwise value, in the case of a grid), piecewise polynomial approximations (splines functions), but other choices are possible. It is desirable to keep the parameter in the "functional" form as long as possible, so that it is possible to easily change parametrization. Of course, this implies that the optimization algorithm must work in the infinite dimension setting. It is possible to formulate the quasi-Newton methods (whose principle are recalled in section A2.3) and study their convergence, in the context of Hilbert spaces.

We will see later that we can generally compute the necessary quantities for the optimization (cost function and gradient) with respect to the discretization parameters. Another parametrization will just be an extra step distinct from the simulation, whose gradient is obtained by application of the chain rule for differentiation. This method allows the separation of the various components of the software: simulation, gradient computation, parametrization and optimization are separated modules with well-specified interfaces. We will return to this topic in section 6.4 and we will discuss examples in the following chapter.

Optimization methods are surveyed in Appendix 2. We now give details on the technical, but important, question of computing the gradient of the error functional, with a particular emphasis on the adjoint state method.

6.3. Methods for computing the gradient – the adjoint state method

In this section, we present several ways to compute the gradient of the cost function [6.10].

The difficulty clearly lies in the computation of the derivative of the (implicitly defined) mapping $a \to u_a$. In this section, we present several methods to compute this gradient. We expose herein the general method. In the following chapter, we will see several explicit examples to explain how this computation can be achieved in a specific situation.

We will first address the finite difference method (it is not recommended, but may still be helpful for providing reference values) and then the sensitivity method. We will insist on the adjoint state method that allows us to carry out this computation for a cost independent of the number of parameters. Which of the latter two methods is best suited to a particular situation depends on the number of parameters with respect to the number of measurements. Table 6.1 presents a comparison of the two methods.

6.3.1. *The finite difference method*

In appearance, this method is very simple to implement, which may explain its popularity, but it is not advisable since its cost is not only proportional to the number of parameters to be identified, but it gives only an approximate result, with an accuracy that is difficult to evaluate. In certain circumstances, it can be used to validate a gradient computation by one of the other methods (sensitivity or adjoint state).

The computation of a partial derivative is replaced by the differences quotient:

$$\frac{\partial J}{\partial a_j} \approx \frac{J(a+h_j)-J(a)}{h_j}. \qquad [6.20]$$

It can immediately be observed that the number of evaluations in J is equal to the number of parameters to be identified (plus one, but this calculation is always necessary). Compared with the sensitivity function method (see section 6.3.2), the cost is equivalent, but only the gradient is obtained, not the Jacobian matrix. In addition, the result is not exact. In finite precision, the error comprises two terms: the approximation error and the round-off error. We will precisely analyze how to combine these two effects. To simplify, we only consider the case of a real valued function of a single real variable $x \to f(x)$.

In this case, the first-order Taylor expansion of f yields:

$$f'(x) = \frac{f(x+h) - f(x)}{h} + h/2 f''(x) + O(h^2). \qquad [6.21]$$

Moreover, suppose that f is computed with a relative accuracy ε_f. This may just be the round-off error, in which case ε_f is the arithmetic precision of the computer (of the order of 10^{-16} in double precision), or a far larger value if f is the result of a complex computation. In this case, what is actually computed is $\tilde{f}(x) = f(x)(1+\varepsilon_f)$ and the difference between the calculated quotient and the real derivative is equal to (ignoring the rounding error due the division):

$$f'(x) - \frac{\tilde{f}(x-h) - \tilde{f}(x)}{h} = h/2 f''(x) + 2\varepsilon_f/h\, f(x) + O(h^2). \qquad [6.22]$$

The sum of the first two terms is minimized by the choice:

$$h = 2\sqrt{\varepsilon_f \left| \frac{f(x)}{f''(x)} \right|}, \qquad [6.23]$$

and the total error is then proportional to $\sqrt{\varepsilon_f}$. In double precision, this means that the derivative will have approximately eight correct digits. Note that the value of the function to be differentiated and of its second derivative have a bearing on the choice of the optimal step, as shown in equation [6.23]. The larger $|f(x)|$ is the larger the value of ε that can be chosen. Similarly, the larger $|f''(x)|$ is, that is the quicker $f'(x)$ varies, the smaller the choice of ε will have to be. Thus, the effective choice of the step remains delicate, even if $\sqrt{\varepsilon_f}$ is a first reasonable estimation.

[DEN 96] contains an algorithm for the choice of the step that takes the various scaling factors into account (in the case of several variables). However, as we advise not to compute the gradient by this method, we will not give any details.

There are still cases where the computation of the gradient by finite differences may be useful: in particular to verify an adjoint state computation. In this case, a parameter is chosen randomly and the gradient is computed at that point by finite differences, for several values of the step (say from 10^{-2} to 10^{-15}). If both computations are correct, the error must pass through a minimum in the neighborhood of 10^{-7}. This point will be illustrated by the example of section 7.3.1 in the following chapter.

6.3.2. *Sensitivity functions*

This is the most natural method to compute the gradient of J. It consists of explicitly diffferentiating the state equation with respect to the parameter a and then using the chain rule. We insist on the fact that this method yields an *exact* result.

In the case of cost functions defined as in [6.10], the gradient and the Hessian assume a simple form. This calculation is detailed in Appendix 2 (see proposition A2.5). The gradient of the functional J is given by the following formula:

$$\nabla J(a) = \Phi'(a)^*(\Phi(a) - d), \tag{6.24}$$

where we recall that Φ is defined in equation [6.4].

To apply this result, a first possibility is therefore to compute the differential of the mapping Φ (in finite dimension, the Jacobian matrix is computed). Since Φ is implicitly defined by the solution of the state equation [6.1] and by the observation equation [6.3], we must resort to the implicit function theorem (more precisely to its corollary that allows the computation of the differential of the implicit mapping once it is known to be differentiable). This result says that the differential of Φ is obtained by differentiating the state equation:

$$\partial_u F(a, u)\delta u + \partial_a F(a, u)\delta a = 0, \tag{6.25}$$

then solving the previous (linear) equation, and by composing with (the derivative of) the observation (we have assumed it was linear):

$$\Phi'(a) = -H \left(\partial_u F(a, u)\right)^{-1} \partial_a F(a, u) \tag{6.26}$$

By combining equations [A2.23] and [6.26], we finally obtain the gradient of J:

$$\nabla J(a) = (\Phi'(a))^* \left(Hu(a) - d_{\text{obs}}\right). \tag{6.27}$$

The main disadvantage of this method lies in the fact that the computation of δu requires the solution of a (linearized) state equation for each value of δa. After passage to finite dimension, this means that the computation of each partial derivative $\partial J/\partial a_j$ requires solving an equation like [6.25]. The cost of computing the gradient is therefore *proportional* to the number of parameters. In many situations of interest, this number can be very large: several hundreds of thousands, or even several millions. We will see in the following (this is the main advantage of the adjoint state method) that it is possible to perform this computation at a cost proportional to that of a single linearized equation and, in particular, *independent* of the number of parameters.

On the other hand, this method provides more than the gradient, since we have seen that it computes the Jacobian matrix of Φ. Once this Jacobian is available, it is possible to make use of it by computing, for example, its singular values. In addition, the Gauss–Newton method requires knowledge of the Jacobian matrix. If the number of parameters is not too high, the Gauss–Newton method can, by computing the gradient as in [6.27], be competitive with a quasi-Newton method with computation of the gradient by the adjoint state.

6.3.3. *The adjoint state method*

We have already noted that the sensitivity functions method provides more than the gradient of J. If we only need the gradient, we can rearrange the computation leading to [6.27] to avoid the full Jacobian computation. By introducing [6.26] in [6.27] and by transposing the product, we obtain:

$$\nabla J(a) = -\Big(\partial_a F(a,u)^* (\partial_u F(a,u)^*)^{-1} H^*\Big)\big(Hu(a) - d_{\text{obs}}\big). \qquad [6.28]$$

The observation, trivial in appearance, which will allow us to simplify the computation, is that it is possible to differently parenthesize this expression:

$$\nabla J(a) = -(\partial_a F(a,u))^* \Big((\partial_u F(a,u)^*)^{-1}\Big(H^*(Hu(a) - \hat{d})\Big)\Big). \qquad [6.29]$$

It is practical to give a name to the quantity inside the second parenthesis and to introduce the vector p as a solution of:

$$\partial_u F(a,u)^* p = -H^*\Big(Hu(a) - d_{\text{obs}}\Big), \qquad [6.30]$$

We call this equation the *adjoint equation*, and p is the *adjoint state*. Once this equation is solved, the gradient is then calculated by:

$$\nabla J(a) = \partial_a F(a,u)^* p. \qquad [6.31]$$

We summarize the results of these computations in a theorem:

THEOREM 6.1.– If p is the solution of the *adjoint equation* [6.30], the gradient of J at point a is given by [6.31], where $u = \Phi(a) = u_a$ is the solution of the state equation [6.1] corresponding to a.

REMARK 6.3.– Theorem 6.1 is of great importance. It provides the general context on which the adjoint state method is based. Nonetheless, it is difficult to apply as such in practice and this is why we propose a simple method to achieve the same result. As a matter of fact, it may be difficult in a practical situation to identify the different adjoints involved, or even the operator F itself. This will be particularly true for evolution problems.

REMARK 6.4.– We can see, according to equation [6.31], that we obtain the full gradient of J by solving the single adjoint equation [6.30]. This means that the adjoint state equation calculates the gradient at a cost proportional to that of the computation of the function itself. In most cases, this cost is a small multiple (between 3 and 5) of the function (this result is demonstrated in [GRI 93]). This justifies, *a posteriori*, the adjoint-state method, and the disuse of the "natural" method met in section 6.3.2.

REMARK 6.5.– The operator that is involved in the adjoint equation is therefore the *adjoint* of $H\partial_u F(a, u)^{-1}$, which is none other than the derivative of the mapping Φ that associates the parameter to the data. This is the property that has given its name to the method. The difference with sensitivity functions computation is that here this transposition is implicit, "hidden" in the definition of the adjoint equation.

REMARK 6.6.– The adjoint equation [6.30] is always linear and this holds even if the state equation was nonlinear. The sources of the adjoint equation are formed by the residual of the observation equation.

In practice, we will see that it is sometimes difficult to carry out the transpositions indicated in theorem 6.1. In the next section, we introduce another way of performing this computation that proves to be simpler in practice.

6.3.4. *Computation of the adjoint state by the Lagrangian*

The previous section showed how to compute the gradient of our cost function by solving only two equations: the state equation, followed by the adjoint equation (the operations of formula [6.31] are typically very simple). As we have already pointed out, the implementation of this result is not always straightforward. We now introduce a technique that leads to the same result, but proves to be more flexible in its usage, as we will see in the examples of the following chapter.

The method is based on what can be seen as a computational trick (but a rigorous justification can be found in the book by Chavent [CHA 09]). It begins with "pretending" that variables a and u vary independently considering the state equation as a constraint. Under these conditions, as we recall in section A2.2, it is natural to introduce a Lagrangian. In our case, this is written, at least formally, since here we do not have a finite number of constraints (the notation $(p, F(a, u))$ is the inner product between these two quantities in the space Z):

$$\mathcal{L}(a,\, u,\, p) = \frac{1}{2}\, \|Hu - d_{\text{obs}}\|^2 + (p,\, F(a,\, u)). \qquad [6.32]$$

The fundamental observation (once again seemingly trivial) is that, if u satisfies the state equation corresponding to the parameter a, we have the identity:

$$\mathcal{L}(a,\, u(a),\, p) = J(a),\ \forall p \in Z.$$

By differentiating this relationship, there follows:

$$J'(a)\delta a = \partial_a \mathcal{L}(a,\, u)\, \delta a + \partial_u \mathcal{L}(a, u)\, \partial_a u(a)\delta a. \qquad [6.33]$$

The difficult part to calculate is $\partial_a u(a)$. If we can *choose* $p \in Z$ so that this term vanishes, we will have a simple expression for the derivative of J. Toward this end, we will consider that $\delta u = \partial_a u(a)\, \delta a$ is an independent quantity and we require that the operator $\delta u \to \partial_u \mathcal{L}(a, u)$ vanishes. We then define the *abstract adjoint equation* by:

$$\partial_u \mathcal{L}(a,\, u(a), p)\, \delta u = 0, \quad \forall \delta u \in U, \qquad [6.34]$$

that is by making the expression of $\partial_u \mathcal{L}$ explicit:

$$(H\delta u,\, Hu(a) - d_{\text{obs}}) - (p,\, \partial_u F(a, u(a))\, \delta u) = 0, \quad \forall \delta u \in U. \qquad [6.35]$$

By identifying with respect to δu, we precisely find once more the adjoint equation [6.30], then the differential of J is calculated by the formula:

$$J'(a)\delta a = (p,\, \partial_a F(a, u(a))\, \delta a), \qquad [6.36]$$

and we can see that in fact we obtain the gradient in J:

$$\nabla J(a) = (\partial_a F(a, u(a)))^* p, \qquad [6.37]$$

which is identical to [6.31].

REMARK 6.7.– In practice, the most useful form of the adjoint equation is the "variational" formulation [6.35]. Indeed, as we have already mentioned, it is not always practical to compute the adjoint operators. On the other hand, it is always simple (we will observe it) to start from form [6.35] and to manipulate it (by integrations or piecewise summations), to result in an explicit adjoint equation.

Similarly, it is often more practical to start from equation [6.36] and to manipulate it to identify the gradient than to use formula [6.37] as such.

This method is sufficiently significant for us to summarize the main steps in algorithm 6.1..

Algorithm 6.1. Practical computation of the gradient of the functional [6.10] by the adjoint state method

1: Define the Lagrangian by [6.32];
2: Solve the adjoint equation [6.35] that determines the adjoint state p;
3: The differential of J is given by [6.36], which allows the gradient of J to be identified.

Table 6.1 compares the gradient computation methods by the sensitivity and the adjoint state methods.

	Sensitivity	Adjoint state
Jacobian computation	Yes	No
Gradient computation	Yes	Yes
Information at each iteration	Gradients, Jacobian (singular values)	Gradients only
Number of systems to solve per time step	Number of parameters	2
Number of parameters	Low	Any
Minimization algorithm	Gauss–Newton (Levenberg–Marquardt)	BFGS
Adaptive parametrization	Difficult	Simple
First-order analysis in the neighborhood of the minimum	Yes	No

Table 6.1. *Comparison of the sensitivity and of the adjoint state method for the computation of the gradients*

6.3.5. *The inner product test*

It is notoriously difficult to validate an adjoint state computation. The most reliable method is based on the definition of the adjoint. Thus, let $b \in Z$ and $c \in U$ be two vectors, and consider the two problems:

$$\partial_u F(a, u)v = b, \tag{6.38}$$

$$\partial_u F(a, u)^* p = c. \tag{6.39}$$

According to the definition of the adjoint, the following equality holds:

$$(\partial_u F(a, u)v,\, p) = (\partial_u F(a, u)^* p,\, v), \tag{6.40}$$

that is to say:

$$(b,\, p) = (c,\, v). \tag{6.41}$$

This last equation is the equation that constitutes the *inner product test*. Its relevance lies in that it only involves the right-hand sides and the solutions of the equations and not the operators themselves. The equality is a consequence of the theory and it is easy to check after programming the adjoint state and the (linearized) direct state. Its numerical accuracy must be at the level of the round-off error. In practice, the right-hand side b is often taken randomly and then we set $c = v$. If the test is satisfied for several different choices of b, we can have confidence in the solution of the adjoint equation.

We illustrate this point in section 7.3.1 in the following chapter.

6.4. Parametrization and general organization

We have now all the elements to describe the general organization of a program aiming to solve an inverse problem of the kind of those described above. We insist on a *modular* organization, for all the reasons usually cited and also for a few specific reasons, in particular to ensure the independence of the gradient computation with regard to the parametrization. We will assume that we have a subprogram that solves the optimization problem. This subprogram will call the *simulator* that solves the direct problem. The simulator calling sequence is determined by the optimization subprogram. To be specific, we assume that the calling sequence is:

```
[f, g] = simul(x),
```

where x represents the current optimization point (a^n in the notation of Appendix 2), f is the value of the cost function at the point x ($J(a^n)$) and g is the (vector) gradient at the same point ($\nabla J(a^n)$). In fact, this subprogram will necessarily be an interface between the optimization routine and the real simulation routine whose calling sequence is necessarily more complex. As a matter of fact, our high-level description skips over a large number of details that should be of concern in a real implementation. For example, the size of the problem (size of vectors x and g) does not explicitly appear. Depending on the language used, this oversight can either be reasonable or not. In Fortran or C, it is necessary to specify the size of the vectors passed in arguments. In languages like C++, and even more so with high-level environments such as MATLAB or Scilab, the vector size can be obtained from the vector itself. A more significant point is that in these instances (and in fact in the vast majority of situations), the simulator requires other data than simply the point x. Keeping in mind the examples discussed above, it can be seen that the geometry of the problem, the time-steps, the sources, etc., must be specified. Once again this aspect, very important in practice, is entirely dependent on the language being used, and falls largely outside of the scope of this book. An elegant proposition for the C++ language has been formulated by Symes and Gockenbach [GOC 99].

As previously mentioned, the *representation* a_h of the parameter to identify in the simulation software will generally be different from the *parametrization* a_{opt} used for the inverse problem. If we follow Chavent's recommendations [CHA 79, CHA 91], we must take the finest representation compatible with the state equation. A consequence is that there is then a parametrization mapping M: $a_{\text{opt}} \mapsto a_h$ that makes it possible to move from the "coarse" parameter used for the optimization to the "fine" representation for the simulation. Moreover, it is often convenient to introduce a second projection mapping P that enforces the constraints of the modeling subprogram that the parameter $M(a_{\text{opt}})$ has to satisfy (e.g. of positivity).

The functional to minimize (what the subprogram `simul` must compute) is then:

$$j(a_{\text{opt}}) = J(P(M(a_{\text{opt}}))) \qquad [6.42]$$

and its gradient is calculated by application of the chain rule. The optimization gradient is given by the post-multiplication of the gradient of J by the *transposed* Jacobian matrix of P, then by that of M. These simple computations are made analytically. The advantage of this approach is that it is simple to change the parametrization, or to add a projection. It is also simple to add a regularization term $1/2\varepsilon^2 \left\|a_{\text{opt}} - a_0\right\|^2$ and to add the term $\varepsilon^2(a_{\text{opt}} - a_0)$ to the gradient. The expensive computation of the gradient by the adjoint state is necessary anyway but is independent of the parametrization.

We can now describe the general structure of the optimization code in the form of a block diagram (Figure 6.1), following Chavent [CHA 91]. Obviously, the box corresponding to "modeling" is where most of the effort goes, both during development and computations.

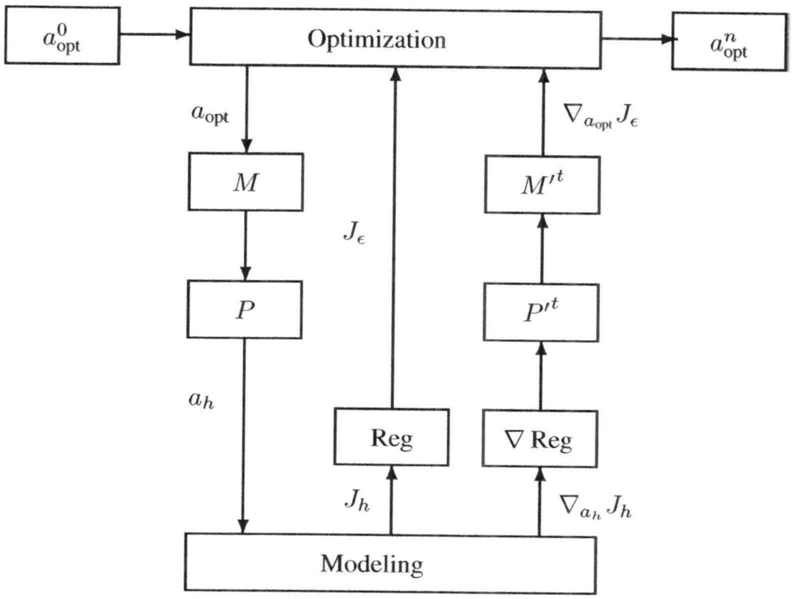

Figure 6.1. *Organization of an inversion code*

6.5. Exercises

EXERCISE 6.1.– About the choice of norms.

Let the matrix be $A = [1\ 1\ 1]^T$ and the vector $b = (b_1, b_2, b_3)^T$ with $b_1 \geq b_2 \geq b_3$. Let $p \in [1, +\infty]$. We want to solve the minimization problem:

$$\min_{x \in \mathbf{R}} \|Ax - b\|_p^p. \qquad [6.43]$$

Compute the solution x for $p = 1, 2, \infty$ (hint: draw a picture). What happens if $b_1 \to +\infty$?

This result gives an indication of the robustness of the norm L^1 with regard to the outlier points.

EXERCISE 6.2.– Identifiability in an elliptic inverse problem.

Consider the boundary problem:

$$\begin{cases} -(au')' = f(x) \quad 0 < x < 1 \\ a(0)u'(0) = 0, \ a(1)u'(1) = 0. \end{cases} \qquad [6.44]$$

The objective is to identify the parameter a from a measurement of u (for simplicity, it is assumed that u is measured over the whole interval $[0, 1]$).

1) We assume:

$$f(x) = \begin{cases} -4x + 1, & 0 < x < 1/2 \\ 0, & 1/2 < x < 1 \end{cases},$$

and

$$u(x) = \begin{cases} x^2 - x + 5/4, & 0 < x < 1/2 \\ 1, & 1/2 < x < 1. \end{cases}$$

Show that for any function φ differentiable in $[1/2, 1]$, with $\varphi'(1/2) = 1$ and $\varphi(1) = 0$, the function

$$a(x) = \begin{cases} x, & 0 < x < 1/2 \\ \varphi(x), & 1/2 < x < 1 \end{cases}$$

satisfies [6.44].

2) Assume now $f(x) = -1$ and $u(x) = x$ for $x \in [0, 1]$. Show that $a(x) = x$.

If u is perturbed to $u_\varepsilon(x) = x + \varepsilon \sin(x/\varepsilon^2)$, check the corresponding coefficient a_ε is $a_\varepsilon(x) = \varepsilon x/(\varepsilon + \cos(x/\varepsilon^2))$.

Show that $\lim_{\varepsilon \to 0} u_\epsilon = u$ uniformly in $[0, 1]$, but that $a_\varepsilon \to 0$ ($\neq a$).

7

Some Parameter Estimation Examples

In this chapter, we apply the theory discussed in the previous chapter to the examples that we have started to examine: the elliptic model in one dimension, then in two dimensions with a distributed parameter, a system of ordinary differential equations and finally the heterogeneous heat equation. We focus on the computation of the gradient of the cost functions that we have defined in the previous chapter, and for some of the models we illustrate the solution of the inverse problem. These examples, of increasing difficulty, will allow us to better understand how to implement the adjoint state method.

7.1. Elliptic equation in one dimension

Recall that the state equation is a boundary value problem in one dimension:

$$\begin{cases} -bu''(x) + cu'(x) = f(x) & 0 < x < 1, \\ u(0) = 0, \ u'(1) = 0. \end{cases} \qquad [7.1]$$

that the observation may take one of the three forms met in example 6.1 and that the corresponding cost functions have been defined by equations [6.12]–[6.14].

We will not address here the issue of the differentiability of the considered mapping. We will reason merely in a formal manner, assuming that the derivatives that we want to compute do exist.

Before starting the computation of the gradient itself, let us give the variational formulation of [7.1]. We define the functions space

$$U = \left\{ s \in H^1(0,1), s(0) = 0 \right\}.$$

The variational formulation of [7.1] is:

Find $u \in U$such that

$$\int_0^1 bu'(x)s'(x) + cu'(x)s(x)\,dx = \int f(x)s(x)\,dx, \quad \forall s \in U. \quad [7.2]$$

The existence and uniqueness of the solution of this problem are consequences of the Lax–Milgram theorem, if b and c are strictly positive. It suffices to verify the coercivity:

$$\int_0^1 bu'(x)^2 + cu'(x)u(x)\,dx = b\int_0^1 u'(x)^2\,dx + cu(1)^2 \geq \|u\|_U^2,$$

according to the Poincaré inequality.

7.1.1. *Computation of the gradient*

This is an example for which a direct computation is simpler than, and as effective as, an adjoint equation computation. We will thus present only this direct computation, keeping aside the adjoint state for the following examples.

We start with the direct computation of the gradient and to illustrate the notions, we will consider the case of $J_1(b,c) = \int_0^1 |u(x) - d_{\text{obs}}|^2\,dx$. We perform a first-order Taylor expansion of J_1 with respect to b and c:

$$\begin{aligned} J_1(b+\delta b, c+\delta c) &= \int_0^1 |u + \delta u(x) - d_{\text{obs}}|^2 \\ &= J_1(b,c) + \int_0^1 (u(x) - d_{\text{obs}})\delta u(x)\,dx, \end{aligned}$$

where δu corresponds to the directional derivative of the mapping $(b,c) \to u$ in the direction $\delta b, \delta c$. In order to calculate this derivative, we can formally differentiate the equation [7.7] with respect to the parameters b and c, and obtain equations whose solutions are the partial derivatives of u with respect to b and c. We obtain the following system, denoting $v = \dfrac{\partial u}{\partial b}$, $w = \dfrac{\partial u}{\partial c}$:

$$\begin{cases} -bv''(x) + cv'(x) = u'' \quad 0 < x < 1 \\ v(0) = 0,\ v'(1) = 0, \\ -bw''(x) + cw'(x) = -u' \quad 0 < x < 1 \\ w(0) = 0,\ w'(1) = 0, \end{cases} \quad [7.3]$$

and $\delta u = v\delta b + w\delta c$.

It follows that

$$J'(b,c)\begin{pmatrix}\delta b\\ \delta c\end{pmatrix} = \int_0^1 (u(x) - d_{\text{obs}}(x))v(x)\,dx\;\delta b + \int_0^1 (u(x) - d_{\text{obs}}(x))w(x)\,dx\;\delta c,$$

which gives the partial derivatives of J:

$$\nabla J_1(b,c) = \begin{pmatrix} \displaystyle\int_0^1 (u(x) - d_{\text{obs}}(x))v(x)\,dx \\ \displaystyle\int_0^1 (u(x) - d_{\text{obs}}(x))w(x)\,dx \end{pmatrix} \qquad [7.4]$$

if v and w are solutions of the linearized system [7.3].

As we have mentioned in section 6.3.2, this computation is not based on any approximation and its cost is proportional to the number of parameters.

For the other two observation cases, we use the same computation of the derivatives v and w, and the gradients are then easily obtained by:

$$\nabla J_2(b,c) = \begin{pmatrix} (u(1) - d_{\text{obs}}(1))v(1) \\ (u(1) - d_{\text{obs}}(1))w(1), \end{pmatrix} \qquad [7.5]$$

and

$$\nabla J_3(b,c) = \begin{pmatrix} (u(1/2) - d_{\text{obs}}(1/2))v(1/2) + (u(1) - d_{\text{obs}}(1))v(1) \\ (u(1/2) - d_{\text{obs}}(1/2))w(1/2) + (u(1) - d_{\text{obs}}(1))w(1) \end{pmatrix}. \qquad [7.6]$$

7.2. Stationary diffusion: elliptic equation in two dimensions

We consider a generalization of the previous example, considered in example 6.3. Given a domain (open-connected subset) $\Omega \subset \mathbf{R}^2$ (to fix ideas), we consider the problem:

$$\begin{cases} -\operatorname{div}(a\,\operatorname{grad} u) = f & \text{in } \Omega \\ u = 0 & \text{on } \Gamma_D \\ a\dfrac{\partial u}{\partial n} = g & \text{on } \Gamma_N \end{cases} \qquad [7.7]$$

where Γ_D and Γ_N form a partition of $\partial\Omega$, $f \in L^2(\Omega)$ and $g \in L^2(\Gamma_N)$ are given.

The parameter to identify is a and we consider here only the cases where u is measured on the Γ_N part of the boundary (since the value of u on $\Gamma_D$is imposed, measuring it would not make any sense). The "natural" space for a is $M = L^\infty(\Omega)$, which will be equipped with the usual L^2 norm (well defined if Ω is bounded)

$$\|a\|_2 = \left(\int_\Omega |a|^2\right)^{1/2},$$

but it should be noted that M is not a Hilbert space for this norm. We will ignore this difficulty in this book. The set of admissible parameters is $M_{\text{ad}} = \{a \in L^\infty(\Omega),\ a \geq a_* > 0 \,\text{ae}\}$ (where a_* is a given constant), for which there is an existence and uniqueness theorem for the direct problem (see below). The cost function is then

$$J(a) = \frac{1}{2}\int_{\Gamma_N} \left|u_{a|\Gamma_N}(x) - d_{\text{obs}}(x)\right|^2 \, d\gamma(x). \qquad [7.8]$$

The adaptation of the method to other situations seen in example 6.3 is easy and left to the reader.

Concerning the direct problem, it is a classical consequence of the Lax–Milgram theorem that if $a \in M_{\text{ad}}$, problem [7.7] admits a unique solution in the space $U = \{u \in H^1(\Omega),\ u = 0 \text{ on } \Gamma_D\}$. The variational formulation of this this problem is

$$\int_\Omega a(x)\,\text{grad}\,u(x)\,\text{grad}\,v(x)\,dx = \int_\Omega f(x)v(x)\,dx + \int_{\Gamma_N} g(x)v(x)\,d\gamma(x),\ \forall v \in U \qquad [7.9]$$

It is possible to apply the theory in the continuous context (see [CHA 09]). However, this calculation is somewhat abstract and since we will eventually have to work in a finite dimensional setting, we will perform the computation after discretization only. Naturally, we will approach the state equation by finite elements that we will choose piecewise linear to simplify. We will assume that the open domain Ω is polygonal, so that we can cover it by a regular triangulation $\mathcal{T}_h$. We will then approximate U by $U_h = \{u_h \in C^0(\overline{\Omega}),\ u_h|_K \in P^1, \forall T \in \mathcal{T}_h,\ u_h = 0 \text{ in } \Gamma_D\} \subset U$. We denote by N_h the number of vertices of the mesh that are not located in Γ_D and by N_Γ the number of vertices in Γ_N.

We must also choose a parametrization for the parameter a. To follow the general method described in section 6.4, we take for the representation of a the finest possible discretization: we will choose a value per triangle (but we will show how to change this parametrization in section 7.2.4). We will therefore denote by a_h the vector representing the discrete function, $a_h = (a_T)_{T \in \mathcal{T}_h}$, and by $M_h = \{a_h, a_T > 0, \forall T \in \mathcal{T}_h\}$ the corresponding space. The discrete representation of the observation is naturally a value per boundary node: d_{obs} will be approximated by the vector $d_h = (d_M,\ M \text{ vertex of } \Gamma_N)$.

It is well known (see a finite elements course; we can cite, for example, [GCC 06, ERN 13] or [DAU 90]) that after discretization, problem [7.7] is reduced to a linear system

$$F(a, u_h) = K(a)\, u_h - L = 0, \qquad [7.10]$$

in which the unknown vector u_h, the stiffness matrix $K(a)$ and the right-hand side L are defined as follows:

– For u_h: We will identify the unknown function $u_h \in U_h$ with the vector of its components in the canonical basis of the finite elements, that is the vector of its values at the vertices of the mesh that are not located on the boundary Γ_D.

– For $K(a)$: The matrix $K(a)$ may be implicitly defined by the equality:

$$\begin{aligned} \forall (u_h, v_h) \in U_h,\ v_h^T K(a) u_h &= \int_\Omega a_h(x)\, \text{grad}\, u_h(x)\, \text{grad}\, v_h(x)\, dx \\ &= \sum_{T \in \mathcal{T}_h} a_T \int_T \text{grad}\, u_h(x)\, \text{grad}\, v_h(x)\, dx. \end{aligned} \qquad [7.11]$$

More explicitly, each element of the matrix $K(a)$ is an integral of the previous kind where the functions u_h and v_h are basis functions of the space U_h. This matrix is usually computed by an assembly procedure (see the references above).

– For L: Finally, the right-hand side of the equation is also calculated by assembly from the equality

$$\forall v_h \in U_h,\ v_h^T L = \int_\Omega f(x) v_h(x)\, dx + \int_{\Gamma_N} g(x) v_h(x)\, d\gamma(x). \qquad [7.12]$$

In general, it is necessary to approximate this integral by a numerical quadrature formula.

Consider now the observation: here, it simply consists of extracting from a vector u_h its components on Γ_N. The observation space D_h is the space of finite element functions on the boundary Γ_N, and we will identify such a function with the vector of its values at the vertices of Γ_N. The discrete observation operator will therefore be represented by a Boolean matrix $H \in \mathbf{R}^{N_h \times N_\Gamma}$ that extracts the degrees of freedom corresponding to the vertices located in Γ_N from the vector $u_h \in U_h$. Note that the adjoint matrix H^T, which we will need below, takes a vector $d_h \in D_h$ and produces a vector in U_h whose elements are zero, except those corresponding to the vertices of Γ_N that have a value equal to that of the element of d_h.

The cost function can be rewritten in the form:

$$\begin{aligned} J(a_h) &= \frac{1}{2}\int_{\Gamma_N} \left|u_h(a_h)(x) - d^h(x)\right|^2 \, d\gamma(x) \\ &= \frac{1}{2}(Hu_h(a_h) - d_h)^T M_N (Hu_h(a_h) - d_h), \end{aligned} \qquad [7.13]$$

where the matrix M_N is defined by

$$\forall (d_h, e_h) \in D_h, \quad e_h^T M_N d_h = \int_{\Gamma_N} d_h \, e_h \, d\gamma(x). \qquad [7.14]$$

This time this is a "mass" matrix, whose elements are the inner products of two basis functions of D_h. It should be noted that this cost function is a norm on D_h but is not the usual Euclidean norm on this space.

Finally, the discrete problem amounts to finding the minimum of the cost function defined by [7.13], $u_h(a_h)$ being the solution of the (linear) state equation [7.10]. To simplify the equations, we will denote simply by u_h the solution $u_h(a_h)$ of the state equation [7.10].

This is a case where we can directly apply the theory of section 6.3.3. But it will also be informative to perform this computation by the Lagrangian method. We will discuss the two methods.

7.2.1. *Computation of the gradient: application of the general method*

In order to use the general method discussed in section 6.3.3, we need to calculate the partial differentials of the mapping F defined in [7.10].

The derivative with respect to u_h is particularly simple, since the mapping F is linear with respect to u_h. It thus yields:

$$\partial_{u_h} F(a_h, u_h)\delta u_h = K(a_h)\delta u_h.$$

F is also linear with respect to a_h, and so we get

$$\partial_{a_h} F(a_h, u_h)\delta a_h = \big(K'(a_h)\delta a_h\big)u_h, \qquad [7.15]$$

but the expression of the differential is complicated to write because the mapping $a_h \rightarrow K(a_h)$ is matrix valued. To compute this differential more easily, it is worthwhile to start again from the variational definition of $K(a_h)$ [7.11]. For two *fixed* vectors (u_h, v_h), we can differentiate the expression [7.11], and we obtain:

$$v_h^T(K'(a_h)\delta a_h)u_h = \sum_{T\in\mathcal{T}_h} \delta a_T \int_T \operatorname{grad} u_h(x) \operatorname{grad} v_h(x)\, dx. \qquad [7.16]$$

According to [6.30], the adjoint equation is written as (note that the adjoint of H is $H^T M_N$ to take the inner product in D_h defined by M_N into account):

$$K(a_h)^T p_h = -H^T M_N\big(Hu_h - d_h\big), \qquad [7.17]$$

where we recall that u_h is a solution of the state equation. The gradient follows by

$$\begin{aligned}
\delta a_h^T \nabla J(a_h) &= \delta a_h^T \partial_{a_h} F(a_h, u_h)^T p_h && \text{by [6.37]}\\
&= p_h^T \partial_{a_h} F(a_h, u_h)\delta a_h\\
&= p_h^T\big(K'(a_h)\delta a_h\big)u_h && \text{according to [7.15]}\\
&= \sum_{T\in\mathcal{T}_h} \delta a_T \int_T \operatorname{grad} u_h(x) \operatorname{grad} p_h(x)\, dx && \text{by [7.16]}.
\end{aligned}$$

The transposition between the first and the second line is permissible because the expression being considered is a scalar. For the last equality, it is important to note that [7.16] has been established in the case where u_h and v_h are vectors that are *independent* of a_h. However, we can now use it with any pair of vectors and the choice of u_h and p_h is then legitimate.

Finally, the partial derivatives of J are given by

$$\begin{aligned}
\frac{\partial J}{\partial a_T} &= \int_T \operatorname{grad} u_h(x) \operatorname{grad} p_h(x)\, dx\\
&= |T|\,(\operatorname{grad} u_h \operatorname{grad} p_h)_{|T}, \quad \forall T \in \mathcal{T}_h,
\end{aligned} \qquad [7.18]$$

because the gradients are constant in each mesh triangle.

It is instructive to interpret the adjoint equation as a variational problem. To this end, we take the inner product of [7.17] with a vector $v_h \in U_h$:

$$v_h^T K(a_h)^T p_h = -v_h^T H^T M(Hu_h - d_h) \qquad [7.19]$$

By transposing this equality, we obtain

$$p_h^T K(a_h) v_h = -(Hu_h - d_h)^T MHv_h, \ \forall v_h \in U_h. \qquad [7.20]$$

We then use [7.11] for the left-hand side and [7.14] for the right-hand side, and we can see that this equality is equivalent to the variational equation (we identify the functions u_h and Hu_h in Γ_N):

$$\int_\Omega a_h(x) \operatorname{grad} v_h(x) \operatorname{grad} p_h(x)\, dx = \int_{\Gamma_N} (u_h(x) - d_h(x)) v_h(x)\, d\gamma(x), \quad \forall v_h \in U_h. \qquad [7.21]$$

It is easy to be convinced that this latter equation is a discretized variational formulation of the partial differential equation

$$\begin{cases} -\operatorname{div}(a \operatorname{grad} p) = 0 & \text{in } \Omega \\ p = 0 & \text{in } \Gamma_D \\ a\dfrac{\partial p}{\partial n} = d_{\text{obs}} - u & \text{in } \Gamma_N \end{cases} \qquad [7.22]$$

As expected, the adjoint equation is linear (but here the state equation was also linear), and its right-hand side originates from the residual of the observation. In this example, the matrix of the adjoint equation is the same as for the state (direct), because the latter is a symmetric matrix.

7.2.2. *Computation of the gradient by the Lagrangian*

We start by forming the Lagrangian:

$$\mathcal{L}(a_h, u_h; p_h) = \frac{1}{2}(Hu_h - d_h)^T M(Hu_h - d_h) + p_h^T K(a_h) u_h, \quad a_h \in M_h, u_h \in U_h, p_h \in U_h$$

The adjoint equation is obtained by differentiating the Lagrangian with respect to the variable u_h. Here, this computation is simple, and the Lagrangian is a quadratic functional with respect to u_h.

$$\partial_u \mathcal{L}(a_h, u_h; p_h) = (Hu_h - d_h)^T MH + p_h^T K(a_h),$$

and thus the adjoint equation [7.17] is found once more.

Similarly, the gradient is obtained by differentiating J_h with respect to a_h:

$$J'(a_h)\delta a_h = p_h^T (K'(a_h)\delta a_h) u_h$$

The computation is identical to the computation that follows [7.16] and, once again, the expression of the gradient with respect to the parameters $(a_K)_{K\in\mathcal{T}_h}$ given in [7.18] is found.

7.2.3. *The inner product test*

In this case, the inner product test takes the following form: given an arbitrary right-hand side b_h:

– solve the direct problem $K(a)\, u_h = b_h$;

– solve the adjoint problem $K(a)^T\, p_h = u_h$;

– verify that $p_h^T b_h = \|u_h\|_2^2$.

To summarize, the adjoint equation involves the *transpose* of the stiffness matrix (which here is symmetrical), then the gradient is easily obtained by equation [7.18].

7.2.4. *Multiscale parametrization*

We illustrate the effect of a change in parametrization, described in general in section 6.4. We take the example of a multiscale parametrization (see [LIU 93]), as we will use this parametrization in the example in section 7.2.5.

The "fine" representation used for the parameter a may not be the most convenient once we get to the optimization stage of solving the inverse problem. So, we assume that, for optimization, we want to represent the parameter a not by using a value per element as previously, but with a different (usually coarser) mesh, comprising n_{opt} parameters. We will assume for simplicity that the computational mesh is a refinement of the optimization mesh, that is to say each optimization cell is constituted of the union of several computational cells. Under these conditions, there is a matrix R: $M_h \to \mathbf{R}^{n_{\text{opt}}}$ such that the new parameter $\tilde{a}$ is obtained from the old parameter a by

$$\tilde{a} = Ra.$$

Matrix R is just a restriction matrix, with R_{Tn} equal to 1 if the element T belongs to the subdomain number n, 0 otherwise. Under these conditions, the cost function and its gradient are simply transformed by

$$\tilde{J}(\tilde{a}) = J(a), \quad \nabla\tilde{J}(\tilde{a}) = R^T\nabla J(a).$$

In practice, starting from an optimization parameter $\tilde{a} \in \mathbf{R}^{n_{\text{opt}}}$, the suitable value is assigned to each element in the triangulation, then the cost function and its gradient with respect to these simulation variables are computed as described in section 7.2.1. Next, this gradient is multiplied by the matrix R^T, which in this case simply amounts to adding all values belonging to the same subdomain.

7.2.5. *Example*

We illustrate the above by an example (in one space dimension). We choose the domain $\Omega =]0,1[$, the diffusion coefficient and the solution

$$a(x) = 1 + x, \quad u_{\text{ex}}(x) = \sin(\pi x), \qquad [7.23]$$

(which yields for the right-hand side $f(x) = \pi^2(1+x)\sin(\pi x) - \pi * \cos(\pi x)$). These two functions are represented in Figure 7.1.

We will take a mesh of the interval $]0,1[$ with $nx + 1$ equal intervals, with $nx = 63$ in the following example. We thus approximate the state by $u_h \in \mathbf{R}^{nx}$ and the parameter by $a_h \in \mathbf{R}^{nx+1}$. The state equation and the adjoint equation have been programmed in MATLAB, resulting in a cost function and its gradient. Concerning the optimization, we have made use of the function `bfgswopt`[1], originating from [KEL 99]. This is an implementation of the BFGS (from the name of its authors Broyden, Fletcher, Goldfarb and Shanno) algorithm (see Appendix 2). In the experiments that follow, the algorithm is initialized with $a_0(x) = 1$.

In accordance with what we have seen in sections 6.4 and 7.2.4, we have separated the representation of the parameter for optimization and discretization. We can thus compare two methods for optimization: a "direct" method, in which we

1 http://www4.ncsu.edu/~ctk/matlab_darts.html.

invert the parameter at a given representation level, and a "hierarchical" method, inspired from [LIU 93], which uses a multiscale parametrization. In this second method, we begin with a scalar parameter a, and at each level, we double the number of parameters, initializing the optimization with the value obtained at the previous level.

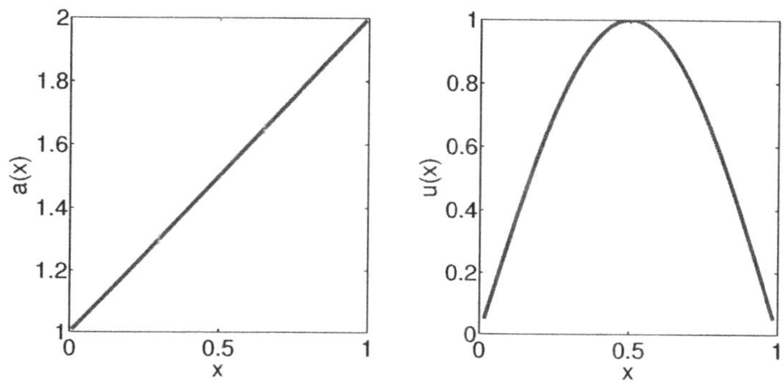

Figure 7.1. *Exact parameter for the inversion (on the left); solution of the direct problem (on the right)*

In a first step, we invert noiseless data. Since these data have been produced by the same code that will be used for the inversion, and on the same mesh, we are committing here an *inverse crime* (see [KAI 05] or [MUÉ 12]). But this will be enough to illustrate our purpose. Figure 7.2 shows the result of the direct optimization with 8 and 64 parameters. As expected (see example 1.2), the parameter is less well determined in the neighborhood of the point where the solution derivative u'_{ex} is zero. Similarly, the results of the hierarchical method are shown in Figure 7.3. It can be observed that the instability in the neighborhood of $x = 1/2$ has been reduced.

Figures 7.4 and 7.5 show the corresponding results when the data are disrupted by random noise, at a level of 1%. The results for eight parameters remain acceptable but those for $nx = 64$ show significant instability. In this case, the hierarchical method does not improve the results.

7.3. Ordinary differential equations

We now examine an example of an evolution problem. We return to example 6.2 and we want to calculate the gradient of the cost function [6.15]. As in the previous section, it is possible to perform this computation on the continuous level but we will only do it after discretization.

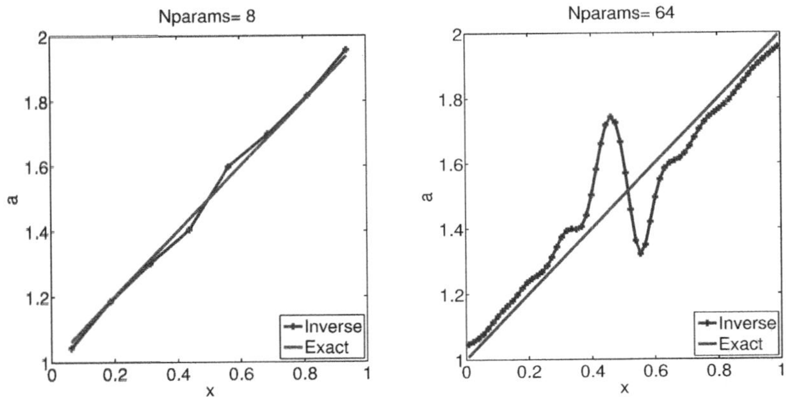

Figure 7.2. *Direct inversion results: on the left, 8 parameters; on the right, 64 parameters. Noiseless data. For a color version of the figure, see www.iste.co.uk/kern/inverse.zip*

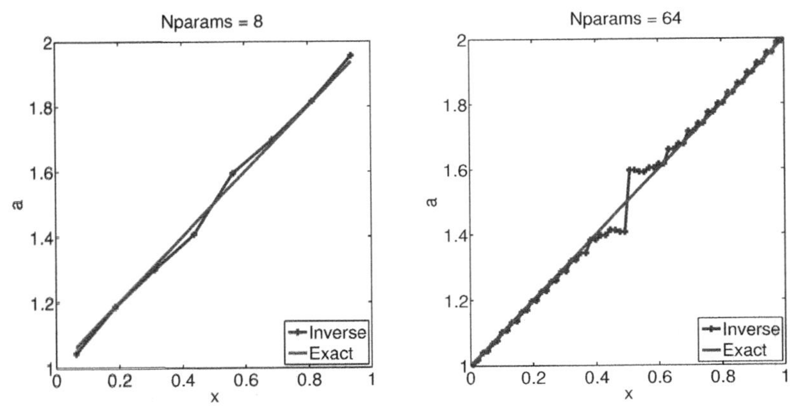

Figure 7.3. *Hierarchical inversion results: on the left, 8 parameters; on the right: 64 parameters. Noiseless data. For a color version of the figure, see www.iste.co.uk/kern/inverse.zip*

The state of the art in numerical methods for ordinary differential equations is very advanced. There are several general software programs of excellent quality (the books [HAI 87] and [HAI 91] constitute a very complete reference). In most situations, it is advisable to use one of these codes rather than writing a simple solver from scratch. We will not follow this advice for the following reasons:

– First, it is preferable to calculate the gradient from the discrete method rather than discretizing a continuous gradient. This would be possible, but in more complex situations than that considered here, it would lead to a different discrete adjoint equation from the one we will obtain below. The advantage of our choice is that it results in the computation of the *exact* gradient of the discrete cost function. Moreover, it avoids having to make choices, that may not always be obvious, to discretize the different quantities (we will demonstrate this in the example of the following section). For a further expanded discussion on this point, see [CHA 79].

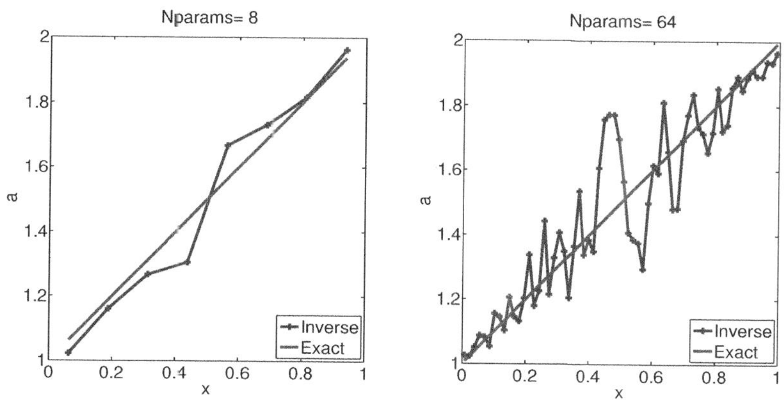

Figure 7.4. *Direct inversion results: on the left, 8 parameters; on the right, 64 parameters. Noisy data (noise =10^{-2}). For a color version of the figure, see www.iste.co.uk/kern/inverse.zip*

– The adjoint equation therefore depends on the integration method used for the direct problem. If this method is itself sophisticated (high order, variable step), it will be more difficult to form the adjoint equation. For this reason, we will confine ourselves to a very simple method, in this case the (explicit) forward Euler method. We nevertheless mention the code CVODES [SER 05] as an example of what can be done to implement an adjoint computation around a differential equations solver that allows an adaptive choice of time steps.

– The other reason is educational. Once the structure of the calculation is understood in the case of the Euler scheme, it will be (in principle) easy to extend the results to more efficient methods.

We thus specify the description of the approximation method of [6.7]. Given $N \in \mathbf{N}$, we let $\Delta t = T/N$, and introduce a subdivision of the interval $[0, T] : 0 = t^0 < t^1 < \cdots < t^N = T$, with $t^n = nh$, $n = 0, \ldots, N$.

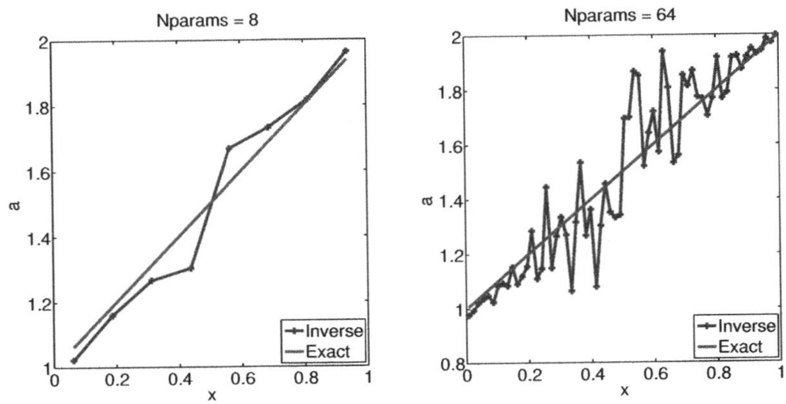

Figure 7.5. *Hierarchical inversion results: on the left, 8 parameters; on the right, 64 parameters. Noisy data (noise = 10^{-2}). For a color version of the figure, see www.iste.co.uk/kern/inverse.zip*

The parameter a is already discrete in this example.

By abuse of notation, we will denote by y^i an approximation of $y(t^i)$. As already mentioned, we choose as approximation schema the forward Euler scheme, defined by

$$\begin{cases} \dfrac{y^n - y^{n-1}}{\Delta t} = f(y^{n-1}, a), \quad n = 1, \dots, N, \\ y^0 = y_0. \end{cases} \qquad [7.24]$$

With regard to the observation, we recall that we have assumed that y was measured at certain times, denoted by $\tau_1, \dots, \tau_Q$. We will make the additional assumption that the time step Δt is chosen such that every instant τ_q is an integer multiple of the time step. This is a very strong simplification, which should be relaxed in a realistic application. This would lead to the introduction of an interpolation operator (and therefore its adjoint), but would not fundamentally change the computation. We will write $n_q = \tau_q / \delta t$, for $q = 1, \dots, Q$.

The cost function has been defined in [6.15] as:

$$J(a) = \frac{1}{2} \sum_{q=1}^{Q} \left| y_a(\tau_q) - d^q_{\text{obs}} \right|^2 \Delta t, \qquad [7.25]$$

where y_a denotes the solution of [7.24], and where we have replaced, in the expression coming from [6.15] Δt_q by Δt.

In this example, it is difficult to directly apply the theory of section 6.3.3, in particular to calculate the transposed derived operator. This will generally be the case for evolution problems. It is with this type of problem that the advantage of the computation by the Lagrangian is the most visible.

This is defined by:

$$\mathcal{L}(a, y, p) = \frac{1}{2}\sum_{q=1}^{Q} |y^{n_q} - d^q|^2 \, \Delta t + \sum_{n=1}^{N} \left(p^{n-1}\right)^T \left(\frac{y^n - y^{n-1}}{\Delta t} + f(y^{n-1}, a)\right) \Delta t. \quad [7.26]$$

We start by forming the adjoint equation from [6.35]:

$$\begin{aligned}\frac{\partial \mathcal{L}}{\partial y}\delta y = &\sum_{n=1}^{N} \left(p^{n-1}\right)^T \left(\frac{\delta y^n - \delta y^{n-1}}{\Delta t} - \partial_y f(y^{n-1}, a)\delta y^{n-1}\right) \Delta t \\ &+ \sum_{q=0}^{Q} (y^{n_q} - d^q)^T \delta y^{n_q} \Delta t = 0, \quad \forall \delta y = (\delta y^0, \dots, \delta y^N).\end{aligned} \quad [7.27]$$

We are going to make this equation more explicit. To this end, we aim to isolate δy^n as a factor. We perform a "discrete integration by parts", that is a simple change in indices:

$$\sum_{n=1}^{N} \delta y^{n-1} p^{n-1} = \sum_{n=0}^{N-1} \delta y^n p^n,$$

which allows us to rewrite the adjoint equation as

$$\begin{aligned}&\sum_{q=1}^{Q} (y^{n_q} - d^q)^T \delta y^{n_q} \, \Delta t + \sum_{n=1}^{N} \left(p^{n-1}\right)^T \delta y^n - \sum_{n=0}^{N-1} \left(p^n\right)^T \delta y^n \\ &\quad - \sum_{n=0}^{N-1} \left(p^n\right)^T \partial_y f(y^n, a)\delta y^n \Delta t = 0, \quad \forall \delta y = \delta y^0, \dots, \delta y^N. \quad [7.28]\end{aligned}$$

We define the vector of residuals r^n, $n = 0, \ldots, N$ by

$$\begin{cases} r^{n_q} = d^q - y^{n_q}, & q = 1, \ldots, Q, \\ r^n = 0 & \text{if } n \text{ is not an observation time.} \end{cases} \qquad [7.29]$$

Since the initial condition does not depend on a, we must have $\delta y^0 = 0$ (more specifically y varies in an affine function space such that $y^0 = y_0$, and δy varies in the associated vector space). Moreover, we can rewrite the last two sums of equation [7.28] as sums from 0 to N, *provided we introduce a new multiplier* p^N, *and we impose* $p^N = 0$. We thus obtain:

$$-\sum_{n=0}^{N} (r^n)^T \delta y^n \Delta t + \sum_{n=0}^{N} \left(p^{n-1}\right)^T \delta y^n - \sum_{n=0}^{N} (p^n)^T \delta y^n$$
$$-\sum_{n=0}^{N} (p^n)^T \partial_y f(y^n, a) \delta y^n \Delta t = 0, \quad \forall \delta y = (\delta y^0, \ldots, \delta y^N). \qquad [7.30]$$

Since this equation is valid for any choice of $\delta y^0, \ldots, \delta y^N$, we can choose all the variations equal to zero except those corresponding to one time step at a time. There follows, after a transposition:

$$\frac{p^{n-1} - p^n}{\Delta t} + \partial_y f(y^n, a)^T p^n = r^n, \quad n = N, \ldots, 1, \qquad [7.31]$$

to which we must add the final condition $p^N = 0$.

The adjoint equation thus appears to be a *backward in time* differences method. Here, the effect of the transposition is to reverse time. As we have already noted, this equation is linear, since y^n is known when calculating p^{n-1}.

We can now calculate the differential with equation [6.36]:

$$J'(a)\,\delta a = \sum_{i=0}^{N} (p^n)^T\, \partial_a f(y^n, a)\, \delta a, \qquad [7.32]$$

and obtain from this the gradient of J:

$$\nabla J(a) = \sum_{n=0}^{N} \partial_a f(y^n, a)^T p^n. \qquad [7.33]$$

REMARK 7.1.– We have seen that equation [7.31] must be integrated backward in time. This observation is a source of difficulty during the computational implementation (not for this very simple example, but this would be the case for more realistic examples, such as that in the following section). In effect, formula [7.33] shows that we need to know y^n and p^{n-1} at the same time to calculate the gradient of J. However, these two quantities are calculated by making "time" evolve in two different directions. In addition, y^n is required for computing p^{n-1} in the adjoint equation. The natural way to compute the gradient is to accumulate the sum in [7.33] at the same time when we calculate p^{n-1}. This means that we need to be able to regenerate the direct state y^n, either by having it stored (possibly on external media), or by recomputing it.

It is clear that neither of these two solutions are satisfactory:

– The first requires storage space proportional to the number of computational stages for the direct state, which is unacceptable for realistic problems (bear in mind that we have in sight the discretization of evolution problems and that in this case, the number of time steps is usually related to the space size of the problem).

– In the general case, the second solution will require an unacceptable computation time. Indeed, the direct problem has no reason to be reversible in time, and therefore the computation of y^n needs to integrate the direct equation up to time n. Similarly, to calculate y^{n-1} the integration must be repeated until time $n-1$, and so on.

An "optimal" compromise has been found by Griewank [GRI 00b, RES 98] and consists of a recovering strategy. Recovery points are determined so as to minimize the number of trajectory recomputations, the available memory for the stores being fixed. Griewank calls this procedure *revolve* and provides an implementation.[2]

We conclude the section with an important addition: the test of the inner product to validate the adjoint state computation. Here again, it is inconvenient to directly apply the general formula [5.41]. It is easier to start from the adjoint equation [7.31]. We take the scalar product of each equation with a vector (that will be specified below) δy^n and we sum all the obtained equations. We obtain:

$$\sum_{n=1}^{N}\Big(\frac{p^{n-1}-p^n}{\Delta t}+\partial_y f(y^n,a)^T p^n\Big)^T \delta y^n=\sum_{n=1}^{N}(r^n)^T\delta y^n. \qquad [7.34]$$

We again perform a "discrete integration by parts" and this yields:

$$\sum_{n=0}^{N-1}\big(\frac{p^n}{\Delta t}\big)^T\delta y^{n+1}+\sum_{n=1}^{N}\Big(-\frac{p^n}{\Delta t}+\partial_y f(y^n,a)^T p^n\Big)^T\delta y^n=\sum_{n=1}^{N}(r^n)^T\delta y^n \qquad [7.35]$$

2 http://www.autodiff.org/?module=Tools&tool=Treeverse%20%2F%20Revolve.

that is (since $p^n = 0$, and by choosing $\delta y^0 = 0$),

$$\sum_{n=0}^{N} (p^n)^T \left(\frac{\delta y^{n+1} - \delta y^n}{\delta t} + \partial_y f(y^n, a) \delta y^n \right) = \sum_{n=1}^{N} (r^n)^T \delta y^n. \qquad [7.36]$$

We take now an arbitrary right-hand side $(b^n)_{n=0,\dots,N}$, and we choose δy^n as a solution of *the linearized state equation*:

$$\frac{\delta y^{n+1} - \delta y^n}{\delta t} + \partial_y f(y^n, a) \delta y^n = b^n, \quad n = 0, \dots, N \qquad [7.37]$$

with the initial condition $\delta y^0 = 0$, equation [7.36] simply becomes:

$$\sum_{n=0}^{N} (p^n)^T b^n = \sum_{n=0}^{N} (r^n)^T \delta y^n. \qquad [7.38]$$

The significance of this equality results from the fact that an adjoint state calculation, whose solution has no physical meaning, is difficult to validate. By controlling equality [7.38] with several randomly selected right-hand sides, we can gain a high confidence level in an adjoint state computation. In order to carry out this computation, it will be necessary to program the solution of the linearized equation, which in our case would not have been necessary. In a more realistic case, at least with an implicit scheme, we would anyway have computed the Jacobian $\partial_y f(y^n, a)$.

It will be helpful to complete this "validation" (which formally is not one) by a verification of the gradient by finite differences.

7.3.1. *An application example*

We illustrate the methods of the previous section on a (simplified) chemical kinetics example: the isomerization of a compound called α-pinene (taken from [DOL 04]). The corresponding differential system is:

$$\begin{cases} y_1' = -(a_1 + a_2) y_1, \\ y_2' = a_1 y_1, \\ y_3' = a_2 y_1 - (a_3 + a_4) y_3 + a_5 y_5, \\ y_4' = a_3 y_3, \\ y_5' = a_3 y_3 - a_5 y_5, \end{cases}$$

with appropriate initial conditions. The state y here represents the different concentrations of the chemical species being present, whereas the parameters a_k are the different reaction rates. The aim is therefore to identify the reaction rates from concentration measurements.

We have implemented (in Scilab, see [KER 04]) a code to solve this inverse problem, based on the adjoint state method. We have first validated the computation of the adjoint state, then that of the gradient according to the method discussed in section 7.3. Defining $a_{\text{ref}} = (1, 1, 1, 1, 1)^T$ and a random right-hand side for the linearized equation, we have obtained, for equation [6.41], the values

$$(p, b) = \sum_{n=0}^{N} (p^n)^T b^n = 442565194.6125011$$

$$(\delta y, \delta y) = \sum_{n=0}^{N} \|\delta y^n\|^2 = 442565194.6125012.$$

The agreement of almost all significant digits is normal and should be observed, insofar as the equality between these two quantities is *exact* and does not depend on the discretization. This is an important ingredient during the development of the inversion computation. If the error exceeds a small multiple of the rounding error (around 10^{-16} in double precision), it is likely that an error remains in the computation of the adjoint state.

To validate the gradient computation, we have calculated the directional derivative of J, around the point a_{ref}, in a fixed direction δ, and for a random observation. The value of this derivative by the adjoint state method is

$$\nabla J(a)^T \delta a = -1.2649927$$

and the result of the computation by finite differences (with perturbations between 10^{-4} and 10^{-12}) is presented in Table 7.1. As expected from the theory (see section 6.3.1), the minimum is located in the neighborhood of $h = 10^{-7}$, and the pattern that can be observed in the column of the differences is consistent with the predictions of the computation (see equation [6.22]).

The values of the parameters used in the example as well as the values obtained after optimization are presented in Table 7.2, whereas Figure 7.6 shows the (synthetic) data, along with the simulated data with the parameters obtained by inversion.

The parameters a_1, a_2, a_3 are recovered rather accurately, while the values of a_4 and a_5 are quite distant from their true values. Nevertheless, these parameters allow the reproduction of the observed data with good accuracy.

h	Derivative	Difference
10^{-4}	−**1264**8470	0.0001457
10^{-5}	−**126497**81	0.0000146
10^{-6}	−**126499**12	0.0000015
10^{-7}	−**1264992**5	0.0000002
10^{-8}	−**1264992**2	0.0000005
10^{-9}	−**126498**99	0.0000028
10^{-10}	−**126499**70	0.0000043
10^{-11}	−**12645**884	0.0004043
10^{-12}	−**125**94370	0.0055557
Adjoint state	−**12649927**	

Table 7.1. *Directional derivative computation by finite differences*

Exact parameters	2.6	1.3	0.9	12.	1.7
Estimated parameters	2.62	1.38	1.00	5.39	0.39

Table 7.2. *Isomerization of α-pinene. Exact and estimated parameters*

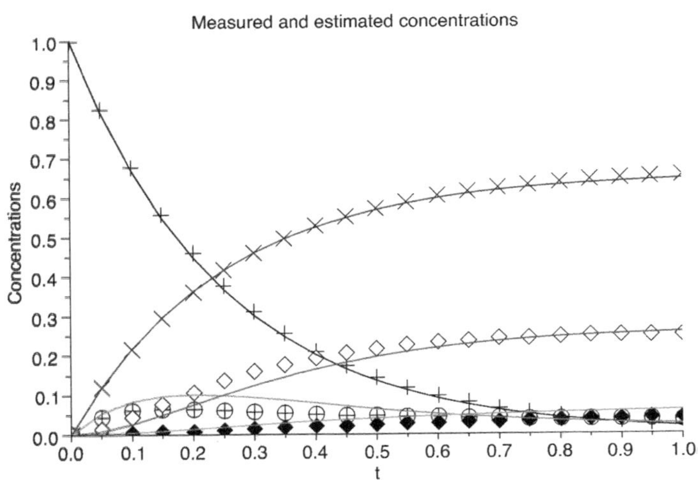

Figure 7.6. *Isomerization of α-pinene. Symbols: data, curves: results after inversion. For a color version of the figure, see www.iste.co.uk/kern/inverse.zip*

In order to explain why some parameters are better estimated than others, a *sensitivity study* should be carried out in this model. Such a study would show how the observations depend on the parameters and how modification of a parameter is reflected in the simulated data. A first approach would be to compute the singular values and singular vectors of the Jacobian matrix of the mapping $F : a \to F(a)$, which indicate the directions that are more likely able to influence measures. An example of this approach, for a different application, can be found in [MAR 08].

This approach is said to be "local" because it is based on linearization and therefore allows us to explore the sensitivity of the parameters only in the neighborhood of a given point. A global approach exists based on a probabilistic treatment of the inverse problem. An introduction may be found in [SAL 08].

7.4. Transient diffusion: heat equation

This example is the synthesis of the previous two: we consider a discretization of the heat equation by finite elements in space and finite differences in time. We follow in this section notes from Chavent [CHA 96]. We start from the variational formulation of the heat equation, before providing a discrete method.

The heat equation is the basic model governing diffusion phenomena and is involved in a number of areas of physics (see equation [2.4]). Given a domain $\Omega \subset \mathbf{R}^2$ (to illustrate the concepts) and a real $T > 0$, we consider the problem:

$$\left\{\begin{array}{rl} \dfrac{\partial u}{\partial t} - \operatorname{div}(a \operatorname{grad} u) = f & \text{in } \Omega\times]0,T[, \\ u(x,t) = 0 & \text{on } \Gamma_D\times]0,T[, \\ a\dfrac{\partial u}{\partial n} = g & \text{on } \Gamma_N\times]0,T[, \\ u(x,0) = u_0(x) & \text{in } \Omega, \end{array}\right. \qquad [7.39]$$

where $f \in L^2(0,T;\ L^2(\Omega))$, $g \in L^2(0,T;\ L^2(\Gamma_N))$, and $u_0 \in L^2(\Omega)$ are given functions (and assumed to be known), and we aim to identify the function a. With the now usual reservations about its non-Hilbertian nature, the natural choice for the space M is the same as in the stationary case in section 7.2: $M = L^\infty(\Omega)$ and $M_{\text{ad}} = \left\{a \in M,\ a(x) \geq a_* > 0\right\}$.

We assume the existence of two types of measurements (see example 6.4): as in section 7.2, we will make the assumption that u is measured over the part Γ_N of the boundary, but also that $u(x,T)$ is known over the entire domain Ω at the final time. Under these conditions, the data consist of two functions: $\hat{d}_N \in L^2(0,T;\ \Gamma_N)$ and $\hat{d}_T \in L^2(\Omega)$. The observations space is then $D = L^2((0,T);\Gamma_N) \times L^2(\Omega)$.

The cost function is

$$J(a) = \frac{1}{2}\int_0^T \int_{\Gamma_N} \left|u - \hat{d}_N\right|^2 \, dx\, dt + \frac{1}{2}\int_\Omega \left|u(x,T) - \hat{d}_T\right|^2 \, dx \qquad [7.40]$$

It is usual to consider [7.39] in a variational form so that we can derive both an existence theorem and the approximation method. It is common to define

$$H = L^2(\Omega), \quad V = \{v \in H^1(\Omega), v = 0 \text{ in } \Gamma_D\}$$

The weak formulation is then

$$\begin{cases} \text{Find } u \in L^2(0,\, T;\, V) \cap C^0(0, T;\, H) \quad \text{such that } \forall v \in V: \\ \dfrac{d}{dt}\displaystyle\int_\Omega u(t)v + \int_\Omega a \,\mathrm{grad}\, u(t)\, \mathrm{grad}\, v = \int_\Omega f(t)v + \int_{\Gamma_N} g(t)v, \quad \text{ae in }]0,\, T[\\ u(0) = u_0 \end{cases}$$

[7.41]

We can state an existence result for problem [7.41]. We refer the reader, for instance, to [DAU 90, vol. 5] for more details and we will directly move to the approximation method. With regard to the discretization in space, we will use a method of finite elements discussed in section 7.2 of which we retain the notations. In addition, we define the mass matrix M_Ω by

$$v_h^T M_\Omega u_h = \int_\Omega u_h v_h.$$

We will choose the Crank–Nicolson method for the time discretization (see [DAU 90, vol. 6] or [ERN 13, KNA 03] for more details of both the finite element method and for methods for evolution problems). We do not assume that the time discretization is necessarily uniform and we will therefore take for a partition $0 = t^0 < t^1 < \cdots < t^N = T$, $\Delta t^{n+1/2} = t^{n+1} - t^n$. We will see that for this choice, the discrete adjoint problem will not necessarily be the same one that would have been obtained by discretization of a continuous adjoint state. We shall also denote by $f_h^{n+1/2}$ and $g_h^{n+1/2}$ approximations to f_h and g_h at time $t^{n+1/2}$. To define the initial condition of the discrete method, we introduce the interpolation operator $I_h : C^0(\Omega) \to U_h$ that associates to a continuous function u_0 the element of U_h (well defined and unique by unisolvence) that takes the same values as u_0 at the mesh points exterior to Γ_D.

With these notations, the direct problem is:

$$\begin{cases} \text{Find } u_h = (u_h^0, u_h^1, \dots, u_h^N) \in U_h^{N+1}, \quad \text{solution of} \\ M_\Omega \dfrac{u_h^{n+1} - u_h^n}{\Delta t^{n+1/2}} + K(a_h)\dfrac{u_h^n + u_h^{n+1}}{2} = M_\Omega f_h^{n+1/2} + M_N g_h^{n+1/2}, \\ u_h^0 = I_h(u_0). \end{cases} \quad [7.42]$$

It should be noted that, in contrast to the previous section, the method defined by [7.42] is implicit, and we must therefore solve a linear system at each time step. The associated matrix is $A^{n+1/2} = M_\Omega + \dfrac{\Delta t^{n+1/2}}{2} K(a_h)$.

We also have to define a discretization of the cost function. The space integrals will be calculated in the same way as in the stationary case in section 7.2. To ensure consistency with the Crank–Nicolson method, we will choose a centered approximation of the time integrals:

$$\int_0^T g(t)\,dt \approx \sum_{n=0}^{N-1} \frac{1}{2}\left(g(t^n) + g(t^{n+})\right)\,\Delta t^{n+1/2}.$$

This leads to a discrete cost function defined by:

$$\begin{aligned} J_h(a_h) = \frac{1}{2}\sum_{n=0}^{N-1} \Big\{ & \frac{1}{2}(u_h^n - \hat{d}_h^n)^T H^T M_N H (u_h^n - \hat{d}_h^n) \\ & + \frac{1}{2}(u_h^{n+1} - \hat{d}_h^{n+1})^T H^T M_N H (u_h^{n+1} - \hat{d}_h^{n+1}) \Big\}\, \Delta t^{n+1/2} \\ + \frac{1}{2}(u_h^N - \hat{d}_T)^T M_\Omega (u_h^N - \hat{d}_T). & \end{aligned}$$

Denoting by $p_h = (p_h^{1/2}, \dots, p_h^{N-1/2})$ the multiplier associated with the equation on the interval $[t^n, t^{n-1}]$, we can then define the discrete Lagrangian:

$$\begin{aligned} \mathcal{L}_h(a_h, u_h, p_h) = & \sum_{n=0}^{N-1} (p_h^{n+1/2})^T \left\{ \frac{u_h^{n+1} - u_h^n}{\Delta t^{n+1/2}} + K(a_h)\frac{u_h^n + u_h^{n+1}}{2} \right\} \Delta t^{n+1/2} \\ & - \sum_{n=0}^{N-1} (p_h^{n+1/2})^T \left\{ M_\Omega f_h^{n+1/2} + H^T M_N g_h^{n+1/2} \right\} \Delta t^{n+1/2} \\ & + \frac{1}{2}\sum_{n=0}^{N-1} \Big\{ \frac{1}{2}(u_h^n - \hat{d}_h^n)^T H^T M_N H (u_h^n - \hat{d}_h^n) \\ & \qquad + \frac{1}{2}(u_h^{n+1} - \hat{d}_h^{n+1})^T H^T M_N H (u_h^{n+1} - \hat{d}_h^{n+1}) \Big\}\, \Delta t^{n+1/2} \\ & + \frac{1}{2}(u_h^N - \hat{d}_T)^T M_\Omega (u_h^N - \hat{d}_T), \end{aligned}$$

Following the usual procedure, we begin by writing the adjoint equation:

$$\frac{\partial \mathcal{L}_h}{\partial u_h}\delta u_h = 0, \quad \forall \delta u_h = (0, \delta u_h^1, \ldots, \delta u_h^N) \in U_h^{N+1}.$$

We therefore obtain, for $\delta u_h \in U_h^{N+1}$:

$$\begin{aligned}
&\sum_{n=0}^{N-1} (p_h^{n+1/2})^T \left\{ M_\Omega \frac{\delta u_h^{n+1} - \delta u_h^n}{\Delta t^{n+1/2}} + K(a_h) \frac{\delta u_h^n + \delta u_h^{n+1}}{2} \right\} \Delta t^{n+1/2} \\
&+ \sum_{n=0}^{N-1} \left\{ \frac{1}{2} \left(u_h^n - \hat{d_h^n} \right)^T H^T M_N H \delta u_h^n \right. \\
&\qquad \left. + \frac{1}{2} \left(u_h^{n+1} - \hat{d_h^{n+1}} \right)^T H^T M_N H \delta u_h^{n+1} \right\} \Delta t^{n+1/2} \\
&+ (u_h^N - \hat{d_T})^T M_\Omega \delta u_h^N = 0
\end{aligned}$$

Next, we perform a discrete integration by parts, that is a shifting of the indices, so as to make δu_h^n appear as factor in all terms. The previous equation becomes, letting $\Delta t^n = 1/2(\Delta t^{n-1/2} + \Delta t^{n+1/2})$

$$\begin{aligned}
&\sum_{n=1}^{N} (p_h^{n-1/2})^T M_\Omega \delta u_h^n - \sum_{n=0}^{N-1} (p_h^{n+1/2})^T M_\Omega \delta u_h^n \\
&+ \frac{1}{2} \sum_{n=1}^{N} (p_h^{n-1/2})^T K(a_h) \delta u_h^n \, \Delta t^{n-1/2} \\
&+ \frac{1}{2} \sum_{n=0}^{N-1} (p_h^{n+1/2})^T K(a_h) \delta u_h^n \, \Delta t^{n+1/2} + \sum_{n=1}^{N-1} (u_h^n - \hat{d_h^n})^T H^T M_N H \delta u_h^n \, \Delta t^n \\
&+ \frac{1}{2} (u_h^N - \hat{d_h^N})^T H^T M_N H \delta u_h^N \, \Delta t^{N+1/2} \\
&+ (u_h^N - \hat{d_T})^T M_\Omega \delta u_h^N = 0, \qquad \forall \delta u_h \in U_h^{N+1}.
\end{aligned} \tag{7.43}$$

Recall that we have $\delta u_h^0 = 0$, which eliminates the terms with $n = 0$ in the previous equality.

In order to interpret the adjoint equation, we operate in two steps: we choose perturbations δu_h of a particular type, acting only on a single instant. We will first address the general case $n < N$, then the final instant $n = N$.

We therefore choose $\delta u_h = (0, \ldots, v_h, \ldots 0)$, where v_h is in position n, for $n = 1, \ldots, N-1$. We obtain, for any $n = 1, \ldots, N-1$:

$$
\begin{aligned}
&(p_h^{n-1/2})^T M_\Omega v_h - (p_h^{n+1/2})^T M_\Omega v_h \\
+ &\frac{\Delta t^{n+1/2}}{2}(p_h^{n-1/2})^T K(a_h) v_h + \frac{\Delta t^{n-1/2}}{2}(p_h^{n+1/2})^T K(a_h) v_h \\
+ &\Delta t^n (u_h^n - \hat{d_h}^n)^T H^T M_N H v_h = 0,
\end{aligned}
$$

that we can rewrite, after transposition, under a form similar to [7.42]:

$$
\begin{aligned}
M_\Omega \frac{p_h^{n-1/2} - p_h^{n+1/2}}{\Delta t^n} &+ K(a_h)\left(\frac{\Delta t^{n-1/2}}{2\Delta t^n} p_h^{n-1/2} + \frac{\Delta t^{n+1/2}}{2\Delta t^n} p_h^{n+1/2}\right) \\
&= -H^T M_N H (u_h^n - \hat{d_h}^n), \quad \forall n = 1, \cdots, N-1. \quad [7.44]
\end{aligned}
$$

There remains a degree of freedom: we choose $\delta u_h = (0, \cdots, v_h)$ in [7.43]. This yields:

$$
\begin{aligned}
&\frac{\Delta t^{N-1/2}}{2}(u_h^N - \hat{d_h}^N)^T H^T M_N H v_h + (u_h^N - \hat{d}_T)^T M_\Omega v_h \\
&\qquad + (p_h^{N-1/2})^T M_\Omega v_h + \frac{1}{2}(p_h^{N-1/2})^T K(a_h) v_h = 0, \quad \forall v_h \in U_h.
\end{aligned}
$$

This equation defines $p_h^{N-1/2}$. We can put it in the same form as for $n < N$, by introducing a fictitious multiplier, defined by

$$
p_h^{N+1/2} = -(u_h^N - \hat{d}_T), \quad \text{and} \quad \Delta t^N = 1/2 \Delta t^{N-1/2}.
$$

REMARK 7.2.– Equation [7.44], including the final condition above, corresponds to a (consistent) discretization of the backward heat equation:

$$
\begin{cases}
-\dfrac{\partial p}{\partial t} - \operatorname{div}(a \operatorname{grad} p) = 0 & \text{in } \Omega \times]0, T[\\
p = 0 & \text{in } \Gamma_D \times]0, T[\\
a \dfrac{\partial p}{\partial n} = -(u - \hat{d}_N) & \text{in } \Gamma_N \times]0, T[\\
p(x, T) = -(u(x, T) - \hat{d}_T) & \text{in } \Omega,
\end{cases}
$$

but the various factors Δt make this discretization non-standard. It is unlikely that we could have obtained this particular method by discretizing the method for the direct problem defined in [7.42].

REMARK 7.3.– Since an observation is taken at the final instant, the final condition on the adjoint state is not homogeneous.

We can finally calculate the gradient of J_h:

$$\delta J_h = \frac{\partial \mathcal{L}}{\partial a_h} = \sum_{n=0}^{N-1} p_h^{n+1/2^T} K(a_h) u_h^{n+1/2} \, \Delta t^{n+1/2}, \tag{7.45}$$

which gives the partial derivative with respect to the elements of a_h:

$$\frac{\partial J_h}{\partial a_T} = |T| \sum_{n=0}^{N-1} \frac{\operatorname{grad} u_{h|T}^n + \operatorname{grad} u_{h|T}^{n+1}}{2} \operatorname{grad} p_{h|T}^{n+1/2} \Delta t^{n+1/2} \tag{7.46}$$

7.5. Exercises

EXERCISE 7.1.– Second-order differential equation.

Consider the (second-order) differential equation:

$$\begin{cases} my''(t) + ky'(t) + \omega^2 y(t) = 0, \quad t \in [0, T] \\ y(0) = 0, \; y'(0) = 1 \end{cases} \tag{7.47}$$

We measure $(y(T), y'(T))$ (the data can be denoted by (d, d')), and the objective is to identify the parameters m, k and ω.

1) Is it possible to identify these three parameters? What can be reasonably expected?

It will be assumed in the following that m is known, and we will note $a = (k, \omega)$, and y_a the solution of [7.47].

2) Give the expression of the cost function J.

3) Show how the partial derivatives of y_a with respect to k and ω can be calculated (they can be denoted, respectively, by v and w.). Derive thereof the expression of the partial derivatives of J with respect to k and ω.

This can either be done directly, or by reduction to a first-order system so as to apply the method discussed in section 7.3.

EXERCISE 7.2.– Identification of a coefficient in the heterogeneous Burgers equation.

Considering a system described by the Burgers equation, generalized to the case of a heterogeneous medium:

$$\begin{aligned}&\frac{\partial u}{\partial t}+\frac{1}{2}\frac{\partial\left(a(x)u^2\right)}{\partial x}=0, \quad 0<x<L,\ t\in[0,T]\\&u(0,t)=0,\\&u(x,0)=u_0(x).\end{aligned} \qquad [7.48]$$

The aim is to identify the coefficient a from the measurements of the solution u at the end of the interval $u(L,t),\ t\in[0,T]$.

1) Formulate the corresponding least squares functional.

2) Compute the gradient when the problem is discretized by the following unisolvence finite difference scheme

$$\begin{aligned}&\frac{u_j^{n+1}-u_j^n}{\Delta t}+\frac{1}{2}\frac{a_j(u_j^n)^2-a_{j-1}(u_{j-1}^n)^2}{\Delta x}=0,\\&\qquad j=1,\cdots,N_x, n=0,\cdots,N_t-1\\&u_o^n=0,\\&u_j^0=u_0(x_j).\end{aligned} \qquad [7.49]$$

EXERCISE 7.3.– Parameter identification in an elliptic problem – theoretical part.

Considering the elliptical problem in dimension 1 according to:

$$\begin{cases}-b\,u''(x)+c\,u'(x)=f(x) & \text{in }]0,1[\\ u(0)=0,\ u'(1)=0\end{cases}, \qquad [7.50]$$

where b and c are real parameters and $f\in L^2(0,1)$.

We seek to identify the coefficients b and c knowing a measurement of $u(x)$ over $]0,1[$.

1) Propose a least squares formulation of this inverse problem.

2) A discretization of the interval $]0, 1[$ by a mesh of step size h is introduced. We define $x_j = jh,\ j = 0, \dots, N,\ u_j \approx u(x_j)$ and $f_j \approx f(x_j)$. An approximation of problem [7.50] by a finite elements method P_1 (u and f are approximated by affine functions on each of the intervals $]x_{j-1}, x_{j+1}[$). Show that the approximated problem is written as:

$$(bK + cD)x = Mb,$$

where $x = (u_1, \dots, u_N)^t$, $b = (f_1, \dots, f_N)^t$. Give the expression of matrices K, D and M.

3) Give the expression of the discrete cost function J.

4) Write the linearized equations and show how to calculate the gradient of the discrete cost function.

5) Write the adjoint equation and show how to calculate the derivative of the cost function.

EXERCISE 7.4.– Parameter identification in an elliptic problem – practical part.

This exercise continues from the previous exercise.

1) Program the solution of the direct problem. To validate, you can use the example ($b = c = 1$): $f(x) = \pi^2/4\ \sin(\pi/2\,x) + \pi/2\ \cos(\pi/2\,x)$ with the solution $u(x) = \sin(\pi/2x)$. Write a function `u=function direct(b,c,f)` where b, c and f are vectors.

2) Program the corresponding functions and calculate the partial derivatives J at the point $b = c = 1$. Validate by finite differences.

3) Program the computation of the adjoint state and of the partial derivatives. Validate the adjoint state by the inner product test and by finite difference derivative.

4) Define f as above. Solve the optimization problem, taking as measurement the solution corresponding to $b = c = 1$, and as initial estimate $b = 2, c = 3$.

Repeat the question by adding a random noise of $1\%, 5\%, 10\%$ to the observations.

In each case, draw the variations of the cost function according to b and c.

8

Further Information

A large number of important topics could not be addressed in this book, for lack of space, lack of knowledge from the author or simply due to subjective choices. This chapter gives some clues on how to address these topics and how to complete those mentioned in the book.

8.1. Regularization in other norms

In Chapter 5, we have used without further ado the norm of the measurements space, which is by hypothesis a Hilbertian norm. This natural choice can be generalized in several directions, which will make it possible to take advantage of the specific properties of the desired solutions.

8.1.1. *Sobolev semi-norms*

In the case of integral operators, the solutions often have regularity properties, which will result in the membership in a Sobolev space. It is then possible to define a regularized problem in the form:

$$J(x) = \frac{1}{2} \|Ax - \hat{z}\|_F + \frac{\varepsilon^2}{2} \|Lx\|_{L^2}, \qquad [8.1]$$

where L is a differential operator (or a discretized version of it) that will most often be the first derivative $L_1 x(t) = x'(t)$ or the second derivative $L_2 x(t) = x''(t)$. After discretization, the differential operator L is replaced by a matrix representing a difference operator.

With an obvious abuse of notation and without imposing boundary conditions, the obtained matrices are as follows:

$$L_1 = \begin{pmatrix} -1 & 1 & & \\ & \ddots & \ddots & \\ & & -1 & 1 \\ & & & 1 \end{pmatrix} \in \mathbf{R}^{(n-1)\times n}$$

for the first derivative, and

$$L_2 = \begin{pmatrix} 1 & -2 & 1 & & \\ & \ddots & \ddots & \ddots & \\ & & 1 & -2 & 1 \end{pmatrix} \in \mathbf{R}^{(n-2)\times n}.$$

for the second derivative.

These two matrices are rectangular and have a non-trivial kernel (corresponding to the constants in the first case, to the affine functions in the second, in agreement with the origin of these matrices).

The theory of Tikhonov regularization in Chapter 5 can be generalized to allow a matrix L, possibly rectangular (which means that the regularization term corresponds to a semi-norm). A condition for the regularized problem to admit a solution is that

$$\ker A \cap \ker L = 0,$$

which is usually satisfied in practice.

This problem falls under the scope of the *general form regularization* theory, for which we mainly refer to Hansen's works. An elementary introduction can be found in Chapter 8 of [HAN 10] and a more complete discussion in [HAN 98], especially concerning the numerical aspects. In addition, these methods are also implemented in the *Regularization Toolbox*. This theory is completed by an extension of the singular value decomposition (*generalized singular value decomposition*), which can also be found in previous references. The more mathematical aspects are described in detail (the expected level is significantly higher than for this book) in [ENG 96].

The significance of the choice of these semi-norms is to select more regular solutions than the Euclidean norm alone would have allowed, since the regularization term penalizes large variations of the derivatives. The abovementioned references contain examples where these choices are helpful and lead to more relevant solutions than the standard L^2-norm regularization.

8.1.2. *Bounded variation regularization norm*

At the other end of the spectrum, applications such as image processing will drive the search for *discontinuous* solutions. Images usually contain well-delimited structures whose sharpest features have to be preserved. Regularizing norms such as those of the previous sections are clearly not suitable.

A choice that has been proved productive is known as *bounded variation* norm, defined (for a regular function in a domain $\Omega \subset \mathbf{R}^d$) by:

$$\|f\|_{BV} = \int_{\Omega} \nabla f| \, dx. \qquad [8.2]$$

This is the L^1 norm of the gradient of the function f. The regularized cost function is then:

$$J(x) = \frac{1}{2} \|Ax - \hat{z}\|_F^2 + \|x\|_{BV} . \qquad [8.3]$$

An important practical difference is that this cost function is no longer quadratic, which will have important consequences on both the mathematical analysis and the implementation of the corresponding method. The first approach, which remains too simple for many applications, consists of approximating the square root by a differentiable function. Thus, instead of [8.3], a differentiable cost function is minimized:

$$J(x) = \frac{1}{2} \|Ax - \hat{z}\|_F^2 - \int_{\Omega} \sqrt{|\nabla x|^2 + \beta^2} \, dx,$$

where $\beta > 0$ is a parameter that obviously remains to be chosen.

This method was introduced in [RUD 92]. An accessible presentation, and pointers to the literature, can be found in Chapter 8 of the book by Vogel [VOG 02] (see also Chapter 7 in [MUE 12]).

8.2. Statistical approach: Bayesian inversion

The approach followed in this book is deterministic, in that it does not make use of any probabilistic notion. It is time to acknowledge that this point of view remains limited and that probability and statistics can bring significant insight to inverse problems.

We will only mention two aspects:

– the link between least squares problems (linear) and statistics;

– some indications about the Bayesian approach to inverse problems.

8.2.1. *Least squares and statistics*

We have often referred to "perturbations" to an exact solution. These perturbations are often linked to experimental uncertainties, and the simplest hypothesis is to assume that the observed errors are random variables. In the "standard linear model", the hypothesis is made that the observations are perturbed by random errors, that the errors are on average zero, are independent and all have the same variance σ^2. In other words, we have a relation of the form:

$$Ax = b = \hat{b} + e,$$

where the vector e is a random vector, with

$$E(e) = 0; \qquad \text{Var}(e) = \sigma^2 I. \qquad [8.4]$$

The solution x of the least squares problem is thus also a random variable. We are going to examine two questions:

– what are the statistical properties of the solution?

– to what extent can we justify the choice of the least squares from a statistical point of view?

With regard to the first question, we will merely give a simple result of the expectation and the variance of x. Note that we do not make any assumption about the error *distribution*, other than those given above in [8.4]. In particular, it is not necessary to assume that the errors follow a Gaussian distribution.

PROPOSITION 8.1.–

$$E(x) = \hat{x} = (A^T A)^{-1} A^T b, \quad \text{Var}(x) = \sigma^2 (A^T A)^{-1}. \qquad [8.5]$$

This result justifies the use of the normal equation [4.3] by statisticians. Its most important consequence is that the variances of the elements of x are given by:

$$\text{Var}(x_i) = ((A^T A)^{-1})_{ii}.$$

The off-diagonal elements of the matrix are the co-variances of the elements of x. It is therefore possible to evaluate *a posteriori* the relevance of the regression model used by computing the variance–co-variance matrix of x.

It should be noted that, contrary to what formula [8.5] seems to indicate, it is not necessary to compute the matrix $(A^T A)^{-1}$. If a QR decomposition of A is available (see Appendix 1), of the form $A = Q \begin{pmatrix} R \\ 0 \end{pmatrix}$ with Q orthogonal and R upper triangular, then

$$\mathrm{Var}(x) = \sigma^2 (R^T R)^{-1} = \sigma^2 R^{-1} R^{-T}.$$

Finally, note that if the variance σ^2 is not known, it can be estimated (unbiased) by the formula:

$$\hat{\sigma}^2 = \frac{1}{m-n} \| Ax - b \|_2^2. \qquad [8.6]$$

Concerning the second question, we now show that the least squares provide the "best" possible estimate of x under the above-mentioned hypotheses. Denote by $\hat{x}$ the solution of the least squares problem corresponding to the "exact" data $\hat{b}$. We find $x = (A^T A)^{-1} A^T (\hat{b} - e)$, and thus by linearity:

$$E(x) = (A^T A)^{-1} A^T E(\hat{b} + e) = (A^T A)^{-1} A^T A \hat{b} = \hat{x}.$$

This property is translated by stating that $\hat{x}$ is an *unbiased* estimator of x.

We thus obtain the following result, whose proof can be found in a statistics course (but also refer to theorem 1.1.1, [BJÖ 96] or theorem 4.32, [VOG 02]).

THEOREM 8.1 (Gauß–Markov).– Among all unbiased linear estimators of x, $\hat{x}$ is the estimator that has the smallest variance.

Thus, any unbiased linear estimator of x has a greater variance than that of $\hat{x}$ It is said that the least squares provide the *best linear unbiased estimate* of x.

In the field of statistics, the above is generally found under the name of regression. We refer to the numerous books on the topic (we recommend the book Aster *et al.* [AST 05], other references to follow in direct connection with our subject are [FAR 88, SAP 90, SEA 71, SEB 77]).

8.2.2. *Bayesian inversion*

8.2.2.1. *A priori and a posteriori probabilities*

The Bayesian approach starts from an interpretation of the probabilities to take into account our lack of information on the system being studied. The parameters of an inverse problem are therefore represented as probability densities, and this enables the introduction of *a priori information* about these parameters. This step comprises a share of subjectivity, which is an integral part of the Bayesian approach. But this subjectivity is explicit here.

We will follow the presentation of Chapter 3 from [KAI 05]. We recall the Bayes formula, a simple result on conditional probabilities. If P is a probability on a probability space Ω, and $(A, B) \in \Omega^2$ are two events:

$$P(A|B) = \frac{P(B|A)\,P(A)}{P(B)}, \quad (P(B) \neq 0), \qquad [8.7]$$

where $P(A|B)$ denotes the *conditional probability* of A knowing B.

In the applications to inverse problems, a probability distribution on the parameter space is therefore introduced, and it is assumed that this distribution admits a density, called *a priori* density $\pi_{\text{pr}}(m)$. The observation, which as in the earlier section, will be perturbed by a random error is also modeled by a random variable, that is assumed to be known in principle; we denote by $\pi(a, d)$ the joint density of (a, d), and we obtain the marginal density

$$\int_D \pi(a, d)\, dd = \pi_{\text{pr}}(a).$$

Naturally, a and d are related by a state equation $d = \Phi(a)$, as mentioned in Chapter 6.

If the value of the parameter a was known, the conditional probability could be derived; it is called *likelihood* (this is a conventional probabilistic concept), and represents the likelihood of possible observations, when the parameter is known:

$$\pi(d|a) = \frac{\pi(a, d)}{\pi_{\text{pr}}(a)}.$$

But naturally, we are faced with a situation where it is the observation that is known. There are measurements available, that is d_{obs}, and the application of the Bayes formula (extended to densities) gives us the *a posteriori density* of a:

$$\pi_{\text{post}}(a) = \frac{\pi_{\text{pr}}(a)\,\pi(d_{\text{obs}}|a)}{\pi(d_{\text{obs}})}. \qquad [8.8]$$

The Bayesian formulation of the inverse problem is therefore:

given an observation d_{obs}, find the *a posteriori* probability $\pi_{\text{post}}(a)$ on the parameters.

In the Bayes formula [8.8], we have implicitly assumed that the denominator was non-zero. This denominator appears as a simple normalization factor (that must ensure that $\int_M \pi_{\text{pr}}(a)\,da = 1$), and in practice the significant quantity is the product of the *a priori* probability $\pi_{\text{pr}}(a)$ by the likelihood $\pi(d_{\text{obs}}|a)$.

Also note that a density known as an *a priori* non-informative density is often utilized, that is to say a uniform density over the parameter space, which may not literally be a probability (e.g. if the parameter space is not bounded). This does not constitute a practical limitation.

Naturally, there exists a relation with the approach mentioned in the previous section. The case of an additive measurement noise, independent of the parameter, results in a probability density of the form

$$\pi(d_{\text{obs}}|a) = \pi_{\text{noise}}(d_{\text{obs}} - \Phi(a)) = \pi_{\text{noise}}(d_{\text{obs}} - Hu_a), \qquad [8.9]$$

where we have introduced the state u_a and the observation operator H defined in Chapter 6. Finally, the *a posteriori* density is defined, up to a normalization constant, by

$$\pi_{\text{post}}(a) = \pi_{\text{pr}}(a)\,\pi_{\text{noise}}(d_{\text{obs}} - Hu_a) \qquad [8.10]$$

The *frequentist approach* to inverse problems consists simply of searching for the parameter a that achieves the *maximum of likelihood*, therefore without introducing any *a priori* probability on the parameter. In the Bayesian approach, this likelihood is modulated by the *a priori* probability on a. Conversely, given the *a priori* probability, $\pi_{\text{pr}}(a)$ is restricted by the observations.

8.2.2.2. *A few estimation techniques*

The question is now to make use of the *a posteriori* density. We will only give a brief discussion, referring to the references in the last section for details, which fall beyond the scope of the study.

The simplest possibility is to search for the *maximum a posteriori* m_{MAP}, which is simply the value (assumed to exist and be unique) that maximizes the *a posteriori* density $\pi_{\text{pr}}(a)\,\pi(d_{\text{obs}}|a)$.

This calculation is relatively simple when considering the case of linear models, and if we make the additional hypothesis that the *a priori* density and the likelihood are both Gaussian, with known variance matrices. It is remarkable that in this (very special) case one recovers once more a generalized least squares problem. The obtained solution is interpreted as the solution of a least squares problem with a Tikhonov regularization, with weighting by the co-variance matrices, and a systematic determination of the regularization parameter that originates from the *a priori* density (see section 3.4 of [KAI 05] for a detailed presentation). This approach can be extended to nonlinear models, where the minimization problem also becomes nonlinear and the state adjoint approach will be useful to calculate the gradient.

However, the Bayesian approach allows a much richer usage, by exploring the *a posteriori* probability. This proposition immediately faces a problem, which is that the parameters space is usually of (very) high dimension. A recent solution is provided by the Markov Chain Monte Carlo methods. Summarized in one sentence, these methods create a Markov chain (sequence of "memoryless" random variables) following (in the limit) the probability density $\pi_{\text{pr}}(m)$. The method remains expensive and it is interesting to note that it has been made possible only due to the increase in computing power. We will not go into the details of these methods, instead referring to the references on this topic below.

8.2.2.3. *References*

The Bayesian approach has been the subject of many publications in recent years. One of the pioneers for the application to inverse problems is Tarantola, whose book [TAR 05] is also available online[1]. The book by Kaipio and Sommersalo [KAI 05] addresses all the aspects of the Bayesian inversion in a more general manner, and the book by Calvetti and Somersalo [CAL 07] offers a presentation close to linear algebra. It is also possible to refer to the book by Idier [IDI 08].

Readers looking for a mathematical approach in a functional context can refer to the works of Stuart, see, for example, [STU 10] and [DAS 16] (of a high mathematical level).

1 http://www.ipgp.fr/~tarantola/Files/Professional/Books/InverseProblemTheory.pdf.

There are many articles to motivate and explain the contribution of the Bayesian approach. One of the most influential references, in the field of geophysics, is [SCA 97] (see also [SCA 01, TAR 00]). The extension to nonlinear problems involves the optimization of a general functional. Examples of the contribution of the Bayesian approach, combined with the power of a supercomputer, can be found in [SCH 12], [BUI 15], BUI 13] and [PET 14].

8.3. Other topics

Briefly, we give some indications about topics that we could not address.

8.3.1. *Theoretical aspects: identifiability*

We have (voluntarily) left aside the theoretical issues related to inverse problems. The notion of identifiability is, nonetheless, central to both the theoretical and practical point of view. This has several possible definitions as follows:

– algebraic identifiability corresponds to the injectivity of the mapping Φ. This notion is difficult to use in practice and does not take account of measurement errors;

– a more manageable definition has been proposed by Chavent [CHA 87]: the parameter a is said to be *identifiable in the sense of least squares* for the measurements $d_{\mathbf{obs}}$ if the functional J admits a solution continuously dependent on $d_{\mathbf{obs}}$. The above-mentioned reference contains a detailed discussion of the practical significance of this condition.

The following references only scratch the surface of a subject with large ramifications: [BAN 89, CHA 02, CHA 09, CHO 09, DUC 13, ISA 98, KUN 88].

8.3.2. *Algorithmic differentiation*

As we have seen in Chapters 6 and 7, while the adjoint state method is conceptually simple, it is well known that it is difficult to program, and especially to validate. Then, why not delegate this work to a program? This is what algorithmic differentiation aims to achieve. Any algorithmic differentiation tool takes as input the source code of a computer program, as well as a list of variables with respect to which the result of this program is to be differentiated. On output, it generates another program that computes, in addition to the results of the original code, the derivatives with respect to the indicated variables. The technique is described in detail in the book by Griewank [GRI 00a]. It can be shown that the inverse mode of algorithmic differentiation is identical to the computation by the adjoint state method [GIL 91].

There are still very few studies on the use of such tools in the context of inverse problems, but meteorologists commonly use them in the context of variational data assimilation. One difficulty, which exists if the adjoint state is computed by hand, is the management of the space–time compromise: as the adjoint state must be computed in backward time, it will be necessary to either store all of the evolution of the direct state, or recompute the state at every time step (we have discussed this point in section 7.3). In addition, algorithmic differentiation programs are forced to take cautious decisions and usually a significant memory usage can be observed. These tools are constantly improving and this is certainly a promising area. Spectacular results have been obtained by combining algorithmic differentiation (over one time step) and "optimal" manual management of the time loop [COL 00, GOC 01, GOC 02].

8.3.3. *Iterative methods and large-scale problems*

We have briefly discussed the regularization of linear inverse problems by iterative methods in section 5.4. These techniques, and their more sophisticated variants, become indispensable when addressing large-size problems for which it is no longer possible to manipulate, or simply store, the matrix of the problem.

Here, the most efficient methods are the conjugate gradient method, or its variants for normal equations. An introduction is presented in [HAN 10], a more complete analysis in [HAN 98] or [BJÖ 96]. More specialized references are [HAN 95], [HAN 93a] and [PAI 82].

Large-scale parameter estimation problems have become affordable only recently due to the development of parallel computing and supercomputers that made possible the use of the Markov Chain Monte Carlo methods mentioned before. We are addressing here a very specialized field, so we provide only a few references: [EPA 08, MAR 12, MON 15, MÉT 13, SAI 12, SCH 12].

8.3.4. *Software*

We conclude this chapter with a short list of software codes and libraries.

We have already mentioned, and used several times in examples, the *Regularization Toolbox*[2] from Hansen [HAN 07]. This tool combines various functions for the regularization of linear problems, mainly from the explicit construction of the matrix SVD. The same author has also published other software programs centered on regularization for more specialized situations. We can cite the following:

2 http://www.mathworks.com/matlabcentral/fileexchange/52-regtools.

– AIR Tools A MATLAB *Package of Algebraic Iterative Reconstruction Methods*[3];

– mxTV *Software for Total Variation Image Reconstruction* [4];

– TVReg *Software for 3D Total Variation Regularization*[5];

– HNO Functions A package of functions destined for image processing to accompany the book Hansen *et al.* [HAN 06].[6]

There is, to my knowledge, no equivalent software to tackle nonlinear problems. After reading this book, the reasons should be clear. There are a few programs specialized in a particular field. One example is PySIT, a toolbox for inversion and seismic imaging[7]. Moreover, geophysics is one of the areas in which a large number of software programs and specialized libraries do exist.

Within more generic software programs, we will simply cite the CVODES / IDAS code [CAO 03], which is a part of the suite SUNDIALS[8] (*SUite of Nonlinear and DIfferential/ALgebraic equation Solvers*) and that proposes code for solving differential and algebraic–differential equations systems, as well as associated adjoint systems.

Software such as OpenTURNS[9] or Dakota[10] [ADA 13] are specialized for the processing of uncertainties, with a strong statistical orientation.

3 http://www2.compute.dtu.dk/~pcha/AIRtools.

4 http://www2.compute.dtu.dk/~pcha/mxTV.

5 http://www2.compute.dtu.dk/~pcha/TVReg/index.html.

6 http://www2.compute.dtu.dk/~pcha/HNO/.

7 http://pysit.bitbucket.org/.

8 https://computation.llnl.gov/casc/sundials/main.html.

9 http://www.openturns.org/.

10 https://dakota.sandia.gov/.

Appendices

Appendix 1

Numerical Methods for Least Squares Problems

In this appendix, we provide a brief review of numerical methods for solving (finite-dimensional) least squares problems on a computer. We will therefore consider problems in finite dimension only (which may possibly originate from the discretization of an integral operator, as we have seen in section 3.2). We therefore consider a matrix $A \in \mathbf{R}^{m \times n}$ and we will assume that $m \geq n$ (the problem is overdetermined). Finally, we will assume, in almost all of this appendix, that matrix A is of maximum rank (i.e. n). The general case, where A is of rank $r < n$, makes use of specific methods. The singular value decomposition (or SVD) can be used in this case, (see sections 4.4.1 and A1.4).

We begin by studying the conditioning of least squares problems. We will then study the conventional method of normal equations that can be solved by Cholesky factorization. We will expose the limitations of this method. We will focus on the method based on a QR factorization of the system matrix, which, although more costly than the previous one, is more robust, and constitutes the method of choice for least squares problems of "reasonable" size (up to a few hundred). Finally, we will give a few (limited) indications about the computational methods for the SVD.

Due to space constraints, we will not discuss specialized methods for large-scale problems, that is the direct methods for sparse matrices, and iterative methods.

We first give some references, where missing details, as well as more advanced material, can be found. Linear algebra is one of the subjects about which there are the highest number of books, and most include a chapter on least squares problems. A good introduction is the book by Ciarlet, either in the English translation [CIA 82] or the French original [CIA 82]. A classic reference is the book by Golub and van

Loan [GOL 96], and an encyclopedic reference on least squares problems is [BJÖ 96]. Other less complete, but more accessible, books are [TRE 97, STE 97] and [DEM 97].

A1.1. Conditioning of the least squares problems

The *condition number* for a rectangular matrix is defined as:

$$\kappa(A) = \frac{\sigma_1}{\sigma_n}, \qquad [A1.1]$$

which (according to proposition 4.5 4) generalizes the usual definition for square matrices.

THEOREM A1.1.– Let $A \in \mathbf{R}^{m\times n}$ be a matrix of rank n, $z \in \mathbf{R}^m$, and let x be a solution of problem [4.2]. Denote $r = Ax - z$ the residual. Finally, let $\tilde{x}$ be a solution of the least squares problem:

$$\min_{x\in\mathbf{R}^n} \|(A+\delta A)\tilde{x} - (z+\delta z)\|^2 .$$

We have the bound

$$\frac{\|\tilde{x}-x\|_2}{\|x\|_2} \lesssim \kappa(A)\left(\frac{\|\delta A\|_2}{\|A\|_2} + \frac{1}{\cos\theta}\frac{\|\delta z\|_2}{\|z\|_2}\right) + \kappa(A)^2\tan\theta\frac{\|\delta A\|_2}{\|A\|_2}, \qquad [A1.2]$$

where $\sin\theta = \dfrac{\|r\|_2}{\|z\|_2}$ and where the symbol $\lesssim$ indicates that this estimation is valid for terms up to order $\max\left(\dfrac{\|\delta A\|_2}{\|A\|_2}, \dfrac{\|\delta z\|_2}{\|z\|_2}\right)$.

PROOF.– We can write $\delta x = \tilde{x} - x$, expand the normal equation for $\tilde{x}$

$$(A^T + \delta A^T)(A+\delta A)\tilde{x} = (A^T + \delta A^T)(z+\delta z)$$

up to first order, and subtract the normal equation for x. It remains (with $r = z - Ax$):

$$A^T A\delta x \approx -A^T\delta Ax + A^T\delta z + \delta A^T r.$$

Taking the norms, we get:

$$\|\delta x\|_2 \lesssim \left\|(A^TA)^{-1}A^T\right\|_2 \|\delta A\|_2 \|x\|_2 + \left\|(A^TA)^{-1}A^T\right\|_2 \|\delta z\|_2 + \left\|(A^TA)^{-1}\right\|_2 \|\delta A\|_2 \|r\|_2 \qquad [A1.3]$$

Denote by $A = U\Sigma V^T$ the SVD of A, with $\Sigma = \begin{pmatrix}\Sigma_1\\0\end{pmatrix}$. Then

$$A^T A = V\,(\Sigma_1\ 0)\,U^T U \begin{pmatrix}\Sigma_1\\0\end{pmatrix} V^T = V\Sigma_1^2 V^T$$

and

$$(A^T A)^{-1} A^T = V\Sigma_1^{-1} U^T.$$

Still according to proposition 4.5 4,

$$\left\|(A^T A)^{-1} A^T\right\|_2 = \frac{1}{\sigma_n} \text{ and } \left\|(A^T A)^{-1}\right\|_2 = \frac{1}{\sigma_n^2}. \qquad \text{[A1.4]}$$

It therefore follows from [A1.3], [A1.4] that

$$\frac{\|\delta x\|_2}{\|x\|_2} \lesssim \kappa(A)\left(\frac{\|\delta A\|_2}{\|A\|_2} + \frac{\|\delta z\|_2}{\|z\|_2}\frac{\|z\|_2}{\|A\|_2\,\|x\|_2}\right) + \kappa(A)^2 \frac{\|\delta A\|_2}{\|A\|_2}\frac{\|r\|_2}{\|A\|_2\,\|x\|_2}.$$

To conclude, we still have to relate the two terms $\frac{\|z\|_2}{\|A\|_2\,\|x\|_2}$ and $\frac{\|r\|_2}{\|A\|_2\,\|x\|_2}$ to the angle θ as stated. However, the residual r is orthogonal to Ax, therefore

$$\cos^2\theta = 1 - \sin^2\theta = \frac{\|z\|_2^2 - \|r\|_2^2}{\|z\|_2^2} = \frac{\|Ax\|_2^2}{\|z\|_2^2}$$

according to the Pythagorean theorem. Finally,

$$\frac{\|z\|_2}{\|A\|_2\,\|x\|_2} \leq \frac{\|z\|_2}{\|Ax\|_2} = \frac{1}{\cos\theta}.$$

□

The angle θ measures the difference between the vectors z and Ax, in other words the size of the residual. It is small if r is small and it is close to $\pi/2$ if r is close to b. It can thus be seen that the conditioning of a least squares problem may depend not on $\kappa(A)$ but on $\kappa(A)^2$. This effect is however mitigated by the presence of the factor $1/\tan\theta$ in front of this term: the effect of $\kappa(A)^2$ will be visible only if the residual is significant.

In detail, if θ is small, the residual is small and the *effective* condition number is close to $\kappa(A)$. If θ is neither small nor close to $\pi/2$, the residual is moderately large, and the effective conditioning is controled by the term in $\kappa(A)^2$, which can be much larger than $\kappa(A)$. Finally, if θ is close to $\pi/2$, the solution is much smaller than the residual, and the effective conditioning becomes unbounded even if $\kappa(A)$ is small.

A1.2. Normal equations

As we have seen in Chapter 4 (equation [4.3]), the solution of a least squares problem amounts to, at least in theory, the solution of a linear system for the matrix A^TA (known as the normal equation matrix):

$$A^TAx = A^Tz. \qquad [A1.5]$$

We have also seen (corollary 4.1) that under the assumption that A is of rank n, the matrix A^TA is defined as positive. This matrix is of size n by n and the normal equations represent information compression, since $n \leq m$. The system [A1.5] therefore can (still in theory) be solved by the Cholesky factorization method. We recall this well-known method:

PROPOSITION A1.1.– Let C be a symmetric positive definite matrix. There exists a unique upper triangular matrix R, with strictly positive diagonal elements, such that

$$C = R^TR. \qquad [A1.6]$$

PROOF.– This can be found in any of the references cited above. □

It then suffices to solve the two linear systems:

$$\begin{cases} R^Ty &= A^Tz \\ Rx &= y \end{cases} \qquad [A1.7]$$

to obtain the solution. The first has a lower triangular matrix; the second an upper triangular matrix.

We also recall in algorithm A1.1 the computational algorithm for the factorization, which consists essentially of identifying, column by column, the elements of R from equation [A1.6].

Note that we have written the algorithm as if matrices C and R occupied different places in memory. In practice, it is usual to let R overwrite C (which is lost).

The number of floating point operations necessary for the formation of the the normal equations matrix is $n(n+1)m + mn$ (by taking advantage of symmetry). The number of operations of the Cholesky factorization is $n^3/3$ floating point operations (additions and multiplications), in addition to n divisions and n extractions of square roots. The solution of the triangular equations [A1.7] takes n^2 operations and can therefore be ignored.

Algorithm A1.1 Cholesky factorization

for $j = 1 : n$ **do**
  for $i = 1 : j - 1$ **do**
$$R_{ij} = \left(C_{ij} - \sum_{k=1}^{i-1} R_{ki} R_{kj} \right) / R_{ii}$$
  end for
$$R_{jj} = \left(C_{jj} - \sum_{k=1}^{j-1} R_{kj}^2 \right)^{1/2}$$
end for

The cost of the normal equations method is therefore, keeping only the main terms: $mn^2 + \frac{1}{3}n^3$.

Despite its simplicity, and its reasonable cost (we will see that other methods for solving least squares problems are more expensive), the normal equations method is not recommended for stability reasons. It passesses two disadvantages:

1) the simple action of forming the matrix $A^T A$ can cause the information on the small coefficients of the matrix A to be lost, which can have disastrous consequences, as shown in example A1.1 below;

2) furthermore, the conditioning of the matrix $A^T A$ is the square of that of A (since the eigenvalues of $A^T A$ are the squares of the singular values of A).

EXAMPLE A1.1 (after Björck [BJÖ 96]).– Consider the problem $\min_{x \in \mathbf{R}^3} \|Ax - z\|$, with

$$A = \begin{pmatrix} 1 & 1 & 1 \\ \varepsilon & 0 & 0 \\ 0 & \varepsilon & 0 \\ 0 & 0 & \varepsilon \end{pmatrix}, \quad z = (1,\ 0,\ 0,\ 0)^T.$$

This example can be a problem where the sum $x_1 + x_2 + x_3$ is known with much greater precision than the components of x individually. The exact computation gives:

$$A = \begin{pmatrix} 1+\varepsilon^2 & 1 & 1 \\ 1 & 1+\varepsilon^2 & 1 \\ 1 & 1 & 1+\varepsilon^2 \end{pmatrix}, \quad A^T z = (1,\ 1,\ 1)^T, \quad x = \frac{1}{3+\varepsilon^2}(1,\ 1,\ 1)^T.$$

Suppose now that $\varepsilon = 10^{-4}$ and that the computation is carried out with eight significant digits. Then, $1 + \varepsilon^2 = 1.000000001$ will be rounded to 1 and the computed matrix $A^T A$ will be singular. Everything behaves as if the last three lines of A were zero and contributed no information.

The difficulties illustrated by this example can be avoided, or at least the boundaries pushed back, by moving into double precision. In these conditions, the normal equations become an acceptable method. However, the method presented in the following section is preferable as a method with more general applicability.

A1.3. QR factorization

The "modern" method to solve least squares problems is based on a factorization known as QR of the matrix A, where Q is an orthogonal matrix, and R is upper triangular. The reduction of A to triangular form is effected by a sequence of multiplications by reflection matrices, known as Householder matrices (in homage to Alston Householder, a pioneer of numerical matrix analysis). We begin by defining and giving the main properties of these matrices before showing how to use them to obtain the QR factorization. We then show how this factorization can be used to solve the least squares problem.

A1.3.1. *Householder matrices*

DEFINITION A1.1.– *Let $u \in \mathbf{R}^n$ be of (Euclidean) norm 1. A Householder matrix is of the form:*

$$H(u) = I - 2uu^T \qquad \text{[A1.8]}$$

Recall that a matrix $Q \in \mathbf{R}^{m\times n}$ is said to be *orthogonal* if it satisfies $Q^TQ = I_n$. Such a matrix preserves the Euclidean norm, $\|Qx\|_2 = \|x\|_2$, and thus corresponds to a linear transformation that is an isometry.

LEMMA A1.1.– A Householder matrix is symmetric and orthogonal.

PROOF.– The symmetry is clear. For the orthogonal character, a simple calculation gives:

$$\begin{aligned}(I - 2uu^T)^T(I - 2uu^T) &= I - 4uu^T + 4(uu^T)(uu^T)\\ &= I - 4uu^T + 4u(u^Tu)u^T) = I,\end{aligned}$$

since $u^Tu = \|u\|_2^2 = 1$. □

From the geometric point of view (see Figure A1.1), a Householder matrix is an orthogonal symmetry relatively to the hyperplane orthogonal to u.

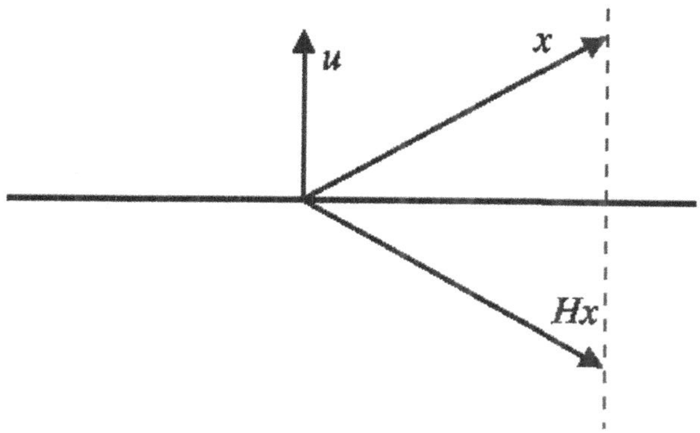

Figure A1.1. *Householder matrix – geometric view*

Note that it is easy to multiply a vector or a matrix by a Householder matrix, knowing only the vector u. Concerning a vector product, it suffices to reconsider the definition [A1.8]:

$$H(u)x = x - 2(uu^T)x = x - 2(u^T x)u, \qquad \text{[A1.9]}$$

which only requires $2n$ operations, instead of n^2 operations if the matrix $H(u)$ is formed and stored. The matrix product is also simple:

$$H(u)A = A - 2(uu^T)A = A - 2u(u^T A), \qquad \text{[A1.10]}$$

only requires $O(n^2)$ operations, which is more economical than to create $H(u)$ and carry out the product.

In our context, the significance of these matrices is that one transforms a given vector into a multiple of the first basis vector.

LEMMA A1.2.– Let $x \in \mathbf{R}^n$ $x \neq 0$. There is a unit vector u such that $H(u)$ is parallel to e_1.

PROOF.– Let $x \neq 0$, we thus look for u such that $H(u)x = \alpha e_1$ (where α is unknown). According to [A1.9], we see that

$$H(u)x = x - 2(uu^T)x = x - 2(u^T x)u$$

lies in the plane spanned by x and u. It is thus possible to search for u with 1-norm, in the plane spanned by x and e_1. Therefore, u must be parallel to $\tilde{u} = x - \alpha e_1$, where

$u = \dfrac{\tilde{u}}{\|\tilde{u}\|_2}$. Finally, α is determined (in absolute value) by $\|H(u)x\|_2 = \|x\|_2 = |\alpha|$, and there are two solution vectors (that are in addition orthogonal). Due to numerical stability reasons, it is preferable to take

$$u = \frac{x + \text{sign}(x_1)e_1}{\|x + \text{sign}(x_1)e_1\|_2}, \qquad \text{[A1.11]}$$

the choice of the sign prevents cancellations during the calculation of the first component of u. □

Now that we have established the tool, let us see how to use it.

A1.3.2. QR *factorization*

THEOREM A1.2.– Let $A \in \mathbf{R}^{m\times n}$ be a rank n matrix. There is an orthogonal matrix $Q \in \mathbf{R}^{m\times m}$ and an invertible upper triangular matrix $R \in \mathbf{R}^{n\times n}$ such that:

$$A = Q\begin{pmatrix} R \\ 0 \end{pmatrix}. \qquad \text{[A1.12]}$$

PROOF.– We give a constructive proof, describing the algorithm being used to obtain the factorization. In this method, matrix Q only appears implicitly, as the product of Householder transformations.

To this end, we proceed in n stages, each of them operating on a column of A by applying a Householder transformation.

The first step consists of determining a Householder matrix that sends the first column of A onto a multiple of the first basis vector. This is precisely the problem that we have solved in the previous section! We will thus define $u_1 = \tilde{u_1} / \|u_1\|_2$ with $\tilde{u_1} = a_1 + \text{sign}(a_{11})e_1$, and we set $P_1 = I - 2u_1u_1^T$. We obtain:

$$A^{(2)} = P_1 A = \begin{pmatrix} r_{11} & r_{12} & \dots & r_{1n} \\ 0 & \tilde{a}_{22} & \dots & \tilde{a}_{2n} \\ \vdots & \vdots & & \vdots \\ 0 & \tilde{a}_{m2} & \dots & \tilde{a}_{mn} \end{pmatrix}.$$

The algorithm then carries on with the submatrix of size $(m-1) \times (n-1)$ located in the lower right corner. Before the k-th step, the reduced matrix has the form:

$$A^{(k)} = \begin{pmatrix} R_{11}^{(k)} & R^{(k)} \\ 0 & \tilde{A}^{(k)} \end{pmatrix}, \qquad \text{[A1.13]}$$

where $R_{11}^{(k)}$ is upper triangular, and will no longer be modified. Then, a Householder matrix $\tilde{P}^k$ is determined that sends the first column of $\tilde{A}^{(k)}$ onto the first basis vector, which gives $u_k = \tilde{u_k} / \|u_k\|_2$ with $\tilde{u}_k = \tilde{a}_k + \text{sign}(a_{kk})e_1$, and we set:

$$P_k = \begin{pmatrix} I_k & 0 \\ 0 & \tilde{P}_k \end{pmatrix}, \qquad A^{(k+1)} = P_k A^{(k)},$$

which is still in the same form.

After n steps, we get the equality:

$$P_n \cdots P_2 P_1 A = \begin{pmatrix} R \\ 0 \end{pmatrix},$$

with R upper triangular, which gives the expected result if we define:

$$Q^T = P_n \cdots P_2 P_1.$$

We conclude by noting that the matrix R is invertible, since we have assumed A of rank n, and that an orthogonal matrix is invertible. □

As we had mentioned, the orthogonal matrix Q is obtained implicitly, as a product of n Householder matrices. It is not necessary (and even useless) to carry out this product. It suffices to preserve the various vectors u_k, $k = 1, \ldots, n$.

We can illustrate the algorithm in the case $m = 6$, $n = 4$. In Figure A1.2, an x represents an element (*a priori* non-zero) of matrix A 0 representing an element certainly equal to zero and an r an element of R that will no longer be modified. Such a figure is called a Wilkinson diagram [TRE 97].

$$\boxed{\begin{matrix} x & x & x & x \\ \hat{x} & x & x & x \\ \hat{x} & x & x & x \\ \hat{x} & x & x & x \\ \hat{x} & x & x & x \\ \hat{x} & x & x & x \end{matrix}} \overset{H_1}{\Longrightarrow} \begin{matrix} r & r & r & r \\ 0 & x & x & x \\ 0 & \hat{x} & x & x \\ 0 & \hat{x} & x & x \\ 0 & \hat{x} & x & x \\ 0 & \hat{x} & x & x \end{matrix} \overset{H_2}{\Longrightarrow} \begin{matrix} r & r & r & r \\ 0 & r & r & r \\ 0 & 0 & x & x \\ 0 & 0 & \hat{x} & x \\ 0 & 0 & \hat{x} & x \\ 0 & 0 & \hat{x} & x \end{matrix} \overset{H_3}{\Longrightarrow}$$

$$\overset{H_3}{\Longrightarrow} \begin{matrix} r & r & r & r \\ 0 & r & r & r \\ 0 & 0 & r & r \\ 0 & 0 & 0 & x \\ 0 & 0 & 0 & \hat{x} \\ 0 & 0 & 0 & \hat{x} \end{matrix} \overset{H_4}{\Longrightarrow} \begin{matrix} r & r & r & r \\ 0 & r & r & r \\ 0 & 0 & r & r \\ 0 & 0 & 0 & r \\ 0 & 0 & 0 & 0 \\ 0 & 0 & 0 & 0 \end{matrix}$$

Figure A1.2. *QR factorization*

Some details of the implementation should be clarified. At each stage, we operate only on the submatrix denoted as $\tilde{A}^{(k)}$ above. In addition, in agreement with what we have observed in the previous section, the product $\tilde{P}_k\tilde{A}^{(k)}$ is calculated by:

$$\tilde{P}_k\tilde{A}^{(k)} = \tilde{A}^{(k)} - 2u_k(u_k^T\tilde{A}^{(k)}).$$

Finally, we give a formal description of the algorithm, see algorithm A1.2.

Algorithm A1.2 QR factorization

for $k = 1, \ldots, n$ **do**
  Determine the vector u_k of the Householder matrix $\tilde{P}_k$ (according to equation [A1.11]).
  for $j = k+1, \ldots, n$ **do**
    $r_{kj} = \tilde{a}_{kj}^{(k)} - u_k^T\tilde{a}_j^{(k)}$
    $\tilde{a}_j^{(k+1)} = \tilde{a}_j^{(k)} - (u_k^T\tilde{a}_j^{(k)})u_k$
  end for
end for

In practice, the matrix R overwrites the upper part of A, and the vectors u_k can be stored instead within the bottom part (the first element of each vector must be stored in an additional array).

It can be shown (see [TRE 97]) that the number of floating point operations is (to first order) $2mn^2 - \frac{2}{3}n^3$. The computation of the QR factorization is thus more expensive than the creation, and the factorization of the matrix of the normal equations. The extra cost is however only of a factor of 2 (the order of complexity is the same). In addition (see the references cited at the beginning of this appendix), the numerical properties of the QR method are superior (due to the fact that we are working with orthogonal matrices), and this makes it the method of choice for a robust implementation. In practice, all numerical libraries provide a version of a solver for the least squares problems based on the QR factorization. In interactive programs like Scilab or MATLAB, the command `x= A\z` solves the least squares problem when A is rectangular, based on the QR factorization of A.

The QR decomposition has an important interpretation in terms of orthogonalization. To clarify this, we partition the matrix Q into $Q = (Q_1\ Q_2)$, with $Q_1 \in \mathbf{R}^{m\times n}$, $Q_1 \in \mathbf{R}^{(m-n)\times n}$. The columns of Q_1 form an orthogonal basis of the image of A, those of Q_2 form an orthonormal basis of $\operatorname{Im} A^\perp$, and we get the reduced factorization:

$$A = Q_1 R. \qquad \text{[A1.14]}$$

This factorization shows that, for $k \leq n$, the first k columns of Q form an orthonormal basis of the subspace generated by the first k columns of A. In addition, as R is invertible, these two subspaces are equal for each k.

Algorithm A1.2 thus builds an orthonormal basis for *all* the subspaces generated by the columns of A. The result obtained is usually found by the Gram–Schmidt orthogonalization. However, there are two notable differences:

– The Gram–Schmidt algorithm is notoriously *unstable*: the vectors calculated by this method are not numerically orthogonal. The QR algorithm computes the same result in a numerically stable manner (see below). It is possible to use a variant (the modified Gram–Schmidt algorithm), which has better stability properties, but those of the QR algorithm are even better.

– The Gram–Schmidt algorithm only calculates the matrix that we have denoted as Q_1. As we have seen, the QR algorithm provides the matrix Q_2 in addition. Depending on the applications, this difference can be significant or not.

REMARK A1.1.– The relation [A1.14] shows that R is none other than the Cholesky factor of A^TA. Indeed, $A^TA = R^TQ_1^TQ_1 = R^TR$, since Q_1 is orthogonal. This observation then shows that R is uniquely determined by A (if A is of rank n), and therefore so is $Q_1 = AR^{-1}$. On the other hand, Q_2 is determined only by the conditions that Q_2 and (Q_1Q_2) be orthogonal, and thus is not unique.

The Householder method presents excellent numerical properties. In the following proposition, we denote by ε_{M} the machine precision (i.e. the smallest positive number such that, on the computer, $1 + \varepsilon_{\mathrm{M}} \neq 1$).

PROPOSITION A1.2.– Let $\overline{R}$ the matrix numerically computed by algorithm A1.2. There is an exactly orthogonal matrix $\tilde{Q}$ (this is *not* the matrix numerically computed) such that:

$$A + E = \tilde{Q}\begin{pmatrix} \overline{R} \\ 0 \end{pmatrix}, \quad \|E\|_2 \leq c(m,n)\varepsilon_{\mathrm{M}}\|A\|_2, \qquad [A1.15]$$

where $c(m, n)$ is a "slowly" increasing function of m and n.

PROOF.– [LAW 95]. □

This result expresses the *backward stability* of the QR algorithm. This concept, introduced by Wilkinson, has proved particularly fruitful in the stability analysis of algorithms. The basic idea is not to try to compare the result of the algorithm (denoted by R) and the computed value (that we denoted $\overline{R}$), which can be very different, but "to reject the error on the data." This is what we have done in proposition A1.2, where $\overline{R}$ is the *exact* R factor of a matrix close to A.

A1.3.3. *Solution of the least squares problem*

Once the QR factorization is obtained, it is easy to solve the least squares problem.

THEOREM A1.3.– Let $A = Q\begin{pmatrix} R \\ 0 \end{pmatrix}$ be the QR decomposition of matrix A. The solution of the least squares problem is given by the solution of the linear system:

$$Rx = z_1, \quad \text{with } Q^T z = \begin{pmatrix} z_1 \\ z_2 \end{pmatrix}. \qquad \text{[A1.16]}$$

PROOF.– Due to the fact that the matrix Q is orthogonal, we get:

$$\|Ax - z\|_2^2 = \left\| Q\begin{pmatrix} R \\ 0 \end{pmatrix} x - z \right\|_2^2 = \left\| \begin{pmatrix} R \\ 0 \end{pmatrix} x - Q^T z \right\|_2^2.$$

Defining $Q^T z = \begin{pmatrix} z_1 \\ z_2 \end{pmatrix}$ as indicated in the statement of the theorem yields:

$$\|Ax - z\|_2^2 = \|Rx - z_1\|_2^2 + \|z_2\|_2^2.$$

The second term is constant (independent of x), and the first is minimized by setting $x = R^{-1} z_1$, which is possible since R is invertible. □

We should add a few words about the effective implementation of the method.

The right-hand side $Q^T b$ is computed implicitly using the representation of Q^T as a product of Householder matrices and [A1.9]:

for $k = 1, \ldots, n$ **do**
  $z_{k:m} = z_{k:m} - 2(u_k^T z_{k:m}) u_k,$
end for

denoting by $z_{k:m}$ the subvector composed of the elements z_k to z_m.

Once this computation is performed, the solution of the triangular system [A1.16] is performed in $O(n^2)$ operations. The main cost of the method is therefore in the initial QR decomposition.

As a consequence of proposition A1.2, we have a result of backward stability for the least squares problem.

PROPOSITION A1.3.– The solution $\overline{x}$ computed in theorem A1.3 is the exact solution of the perturbated problem:

$$\min_{x \in \mathbf{R}^n} \|(A + E)x - (z + f)\|_2, \qquad \text{[A1.17]}$$

where the perturbations E and f satisfy:

$$\|f\|_2 \leq c(m,n)\varepsilon_\mathrm{M} \|z\|_2 \text{ and } \|E\|_2 \leq c(m,n)\varepsilon_\mathrm{M} \|A\|_2 .$$

By combining the results of proposition A1.3 and theorem A1.1, we can estimate the *error* occurring in the solution.

COROLLARY A1.1.– The solution $\overline{x}$ computed in theorem A1.3 satisfies:

$$\frac{\|x-\overline{x}\|_2}{\|x\|_2} \leq c(m,n)\varepsilon_\mathrm{M} \left\{ \kappa(A) \left(\frac{\|\delta A\|_2}{\|A\|_2} + \frac{1}{\cos\theta} \frac{\|\delta z\|_2}{\|z\|_2} \right) + \kappa(A)^2 \tan\theta \frac{\|\delta A\|_2}{\|A\|_2} \right\}, \quad \text{[A1.18]}$$

x being the exact solution of [4.2]

A1.4. SVD and numerical methods

The method that we have seen in section 4.4.1 provides a numerical method with excellent stability properties. The only disadvantage is its cost, higher than the method based on the QR factorization. We recall, in matrix form, the method of section 4.4.1, assuming the singular value factorization of $A = U_n \Sigma V^T$ as known (for simplicity, we limit ourselves to the case where A is of rank n, Σ_1 is therefore $n \times n$, invertible):

Algorithm A1.3 Solution of a least squares problem by SVD

1: Compute $w = U_n^T z$,
2: Solve the (diagonal) system $\Sigma_1 y = w$,
3: Define $x = Vy$.

In order for this method to be effectively implementable, the last point to consider is how to compute the singular value decomposition (or the factorization) of A. This involves more specialized, and different, techniques from those that we have seen in this book and we will only provide general guidance. Details are available in the references cited at the beginning of the appendix (in particular, [DEM 97] and [BJÖ 96]).

Unlike the LU and QR factorizations, a computational algorithm for the SVD is necessarily iterative. This is a consequence of the fact that the singular values of A are the eigenvalues of $A^T A$, that is the roots of the characteristic polynomial of this latter matrix. And there is no *finite* algorithm to compute the roots of a polynomial of degree higher than 5 (this is known since Abel and Galois in the 19th Century).

Moreover, it should be noted that as for normal equations, the method outlined above (compute the eigenvalues of $A^T A$) is *not* advisable.

The most common method for the computation of the SVD operates in two phases:

– First, A is reduced into a bidiagonal form:

$$B = PAQ,$$

where P and Q are two orthogonal matrices (as $B^T B = Q^T A^T AA$, A and B have the same singular values). This computation is done by multiplying A on the left and on the right by Householder matrices and requires a finite number of operations.

– The singular values of B are then computed by a variation of the QR algorithm for tridiagonal matrices. This phase, which is typically iterative, is faster in practice than the previous phase, except if we also want to compute the singular vectors.

Appendix 2

Optimization Refreshers

This chapter presents some optimization results. We refer to some excellent books on the subject (e.g. [BON 97, DEN 96, KEL 99, NOC 99]) for an exhaustive view of the subject. We will limit ourselves to briefly recalling general properties of optimization problems and algorithms, the main features of the BFGS (from the name of its authors Broyden, Fletcher, Goldfarb and Shanno) method, the most important of the quasi-Newton methods, and we will present the Gauss–Newton method, often used for solving nonlinear least squares problems, as well as its variant known as the Levenberg–Marquardt method.

We recall that the problem to be solved is:

$$\min_{a \in M} J(a), \qquad \text{[A2.1]}$$

called *unconstrained minimization*, or the *constrained optimization* variant

$$\min_{a \in M_{\text{ad}}} J(a). \qquad \text{[A2.2]}$$

Throughout this chapter, we will assume that $J : M \to \mathbf{R}$ is a C^2-class mapping from a Hilbert space M to $\mathbf{R}$, which will be sufficient to apply the results that we will need.

A2.1. Local and global algorithms

To solve problem [A2.1], say two major classes of algorithms can be distinguished. Those based on the necessary condition of proposition A2.1 below are qualified as *local*. In practice, this means that these methods make use of the information provided by the gradient of the functional to be minimized and try to solve the equation provided by the *first-order optimality condition* (see proposition A2.1).

The main flaw of this type of algorithm is that it can only converge to a *local* minimum, or even to a critical point (a point where the gradient becomes zero). It is necessary to use the second derivatives to exclude the critical points that are not minima. With only local first-order information, it is not possible to avoid this drawback. It is offset by the fact that these methods are well understood from the mathematical point of view, and that there are quality implementations (a list of existing optimization software programs can be found in [MOR 93]).

The local algorithms are opposed to the methods known as *global* that explore the entire space of parameters so as to converge toward *the* global minimum (if it exists). These methods sometimes have a more heuristic aspect and are inherently slower and more expensive. They have the advantage of avoiding the calculation of the gradient of the cost function, which as we shall see is one of the sensitive points in implementing local methods. Among the global methods, we can cite the *simulated annealing* method [LAA 87], which avoids local minima by allowing the cost function to increase with a certain probability function. We shall also cite the direct search methods [TOR 91, KEL 99], which build a sequence of simplexes that contract toward the global minimum. In both cases, the number of evaluation of the cost function can be very large. Global methods are in fact to be reserved for the cases where a small number of parameters has to be identified, and in which the gradient of the functional is difficult to compute.

In this book, we only consider local algorithms, and we essentially seek to demonstrate how to economically compute the gradient of the cost function.

A2.2. Gradients, Hessians and optimality conditions

For reference, we have gathered in this section a number of definitions and well-known results. The proofs can be found in the above references.

DEFINITION A2.1.–

– *The differential of J at point $a \in M$ is the linear form, denoted $J'(a)$, characterized by the relation:*

$$J(a + \delta a) = J(a) + J'(a)\delta a + o(\|\delta a\|), \quad \forall \delta a \in M \qquad \text{[A2.3]}$$

– *The gradient of J at point $a \in M$, denoted by $\nabla J(a)$, is defined by the equality:*

$$(\nabla J(a),\, h) = J'(a)h, \quad \forall h \in M. \qquad \text{[A2.4]}$$

– *The Hessian of J at point $a \in M$, written $\nabla^2 J(a)$, is the self-adjoint operator on M defined by the equality:*

$$(\nabla^2 J(a)h,\, k) = J''(c)(h,k), \quad \forall (h,\, k) \in M^2. \qquad [A2.5]$$

Moreover, it should be noted that while the differential J' is a topological notion (it is defined by a limit), the gradient is a geometrical notion, since it is related to the choice of an inner product. Similarly, the identification of the bilinear form $J''(a)$ with the Hessian operator depends on the inner product.

In the finite dimensional case, with the usual inner product, the differential is a row vector and the gradient is the associated column vector:

$$J'(a) = \nabla J(a)^T = \left(\frac{\partial J}{\partial a_1}, \ldots, \frac{\partial J}{\partial a_p}\right). \qquad [A2.6]$$

Similarly, the Hessian is identified with a (symmetric) matrix, whose elements are

$$\nabla^2 J(a)_{ij} = \frac{\partial^2 J}{\partial a_i \partial a_j}. \qquad [A2.7]$$

The concept of gradient (and making the distinction with the differential) becomes all the more important in the infinite dimensional case, when there is no longer any canonical identification between a Hilbert space and its dual. Even in the finite dimensional case, it is useful to keep the two concepts, and to rewrite [A2.3] in the form:

$$J(a + \delta a) = J(a) + \nabla J(a)^T \delta a + o(\|\delta a\|), \quad \forall \delta a \in M. \qquad [A2.8]$$

Now recall the relations between gradients, Hessians and the conditions for optimality, first in the unconstrained case.

PROPOSITION A2.1 (Necessary first-order condition).– Let $\hat{a}$ be a point where J reaches its minimum. We have:

$$\nabla J(a) = 0. \qquad [A2.9]$$

This condition is obviously not sufficient, unless J is convex. A critical point of J (a point where ∇J vanishes) can be a minimum, a maximum or even a saddle point. Conversely, we get the following.

PROPOSITION A2.2 (Sufficient second-order condition).– Suppose that $\nabla J(\hat{a}) = 0$ and that $\nabla^2 J(\hat{a})$ is positive definite. Then, J reaches a strict local minimum in $\hat{a}$.

An important notion both in the definition of algorithms and their analysis is as follows.

DEFINITION A2.2.– *A vector $d \in M$ is a descent direction for J at point a if and only if*

$$(\nabla J(a),\, d) < 0 \qquad \text{[A2.10]}$$

This definition ensures that the function decreases, at least locally, along a descent direction. For example, it should be noted that the direction opposite to the gradient is obviously a descent direction. It is on this observation that the steepest descent algorithm is based, but it unfortunately converges too slowly to be useful in practice.

Since this will be helpful when we introduce the adjoint state method, we also provide a basic result in the case of the minimization with equality constraints. We are still trying to minimize a function J defined over a Hilbert space M, but it is assumed that the unknown a must satisfy a constraint of the form $G(a) = 0$, where $G : M \mapsto P$, (P is another Hilbert space) is a mapping (assumed to be differentiable). The following result is known under the name of *Lagrange multipliers theorem*.

PROPOSITION A2.3.– Let $\hat{a}$ a solution of the minimization problem

$$\min_{a \in M} J(a),$$

in addition satisfying $G(\hat{a}) = 0$. Under the hypothesis:

$$G'(\hat{a}) \text{ surjective},$$

there exists $p \in P$, called a *multiplier*, such that:

$$\nabla J(\hat{a}) + G'(\hat{a})^* p = 0 \qquad \text{[A2.11]}$$

In the finite-dimensional case, denote by N_a the dimension of M and by N_p that of P. The constraint G becomes a vector in $\mathbf{R}^{N_p}$ $G = (G_1, \ldots, G_{N_p})$, the same occurs for $p = (p_1, \ldots, p_{N_p})$, and the condition [A2.11] is rewritten as:

$$\nabla J(\hat{a}) + \sum_{j=1}^{N_p} p_j \nabla G_j(\hat{a}) = 0. \qquad \text{[A2.12]}$$

It is usual to introduce the Lagrangian, defined by

$$\mathcal{L}(a,p) = J(a) + \sum_{j=1}^{N_p} p_j G_j(a), \qquad \text{[A2.13]}$$

and equations [A2.11] or [A2.12] express that $\nabla_a \mathcal{L} = 0$, while the constraints follow from $\nabla_p \mathcal{L} = 0$.

A2.3. Quasi-Newton methods

A very effective method to solve [6.10], or rather [A2.9], is Newton's method. It consists of building the sequence a^n defined by solving:

$$H^{n+1}(a^{n+1} - a^n) = -\nabla J(a^n) \qquad \text{[A2.14]}$$

where H^n is the Hessian of J at point a^n. To ensure that the method is well defined, it is obviously important that H^{n+1} be invertible (or better definite positive, which allows to distinguish the minima among the critical points).

It is convenient to define d^n as the solution of the linear system

$$H^{n+1} d^n = -\nabla J(a^n) \qquad \text{[A2.15]}$$

(it is then shown that if H^{n+1} is definite positive, then d^n is a descent direction of J), so that equation [A2.14] is rewritten as:

$$a^{n+1} = a^n + d^n. \qquad \text{[A2.16]}$$

It is well known (see the optimization references above) that Newton's method is locally convergent, with quadratic convergence, provided that the initial point be chosen close enough to the (unknown) solution, and that the Hessian at the solution is not singular. However, it presents rather serious flaws:

– the convergence is not global;

– the algorithm is undefined at points where the Hessian is singular;

– the algorithm does not necessarily generate descent directions;

– the Hessian has to be computed, and the linear system [A2.15] must be solved, at each iteration.

All these flaws are corrected by quasi-Newton methods. We merely briefly expose the BFGS algorithm with linear search, which is one of the most commonly used variant.

The first idea is to keep the descent direction given by [A2.15], but we will not use the whole length of the step to move forward. We replace [A2.16] by:

$$a^{n+1} = a^n + \alpha^n d^n \tag{A2.17}$$

where α^n is determined by a *line search*, that is ensuring that between a^n and a^{n+1} J has sufficiently decreased. It has become clear that it is neither necessary nor even helpful to determine the optimal parameter α^n (which achieves the minimum in α of $J(a^n + \alpha d^n)$). To ensure the convergence of the algorithm, it suffices to satisfy *Wolfe's conditions*:

$$J(a^n + \alpha^n d^n) - J(a^n) \leq \omega_1 \alpha^n (\nabla J(x^n), d^n), \tag{A2.18}$$

$$(\nabla J(a^n + \alpha^n d^n), d^n) \geq \omega_2 (\nabla J(x^n), d^n), \tag{A2.19}$$

where the constants ω_1 and ω_2 satisfy $0 < \omega_1 < \omega_2 < 1$.

The first inequality ensures that J will have decreased by a significant fraction between the previous and the new point. The second simply prevents the step from becoming too small. As a matter of fact, the second condition is not satisfied for $\alpha^n = 0$ if d^n is not a descent direction, and therefore it is still not satisfied for "small" values of α^n (by continuity). Moreover, it can be shown that if d^n is a descent direction and if the function $\alpha \to J(a^n + \alpha d^n)$ is bounded from below, there is a value $\alpha^n > 0$ satisfying the Wolfe conditions. An algorithm that finds a satisfactory value of α is described in [BON 97].

The second modification needed to bring to Newton's method consists of replacing the choice of the direction d^n so as to ensure that it is always a descent direction. The BFGS method solves at once the last three difficulties mentioned above.

This method works by updating in an iterative manner an approximation of the Hessian. It is usual to define the quantities

$$s^n = a^{n+1} - a^n \text{ et } y^n = \nabla J(a^{n+1}) - \nabla J(a^n). \tag{A2.20}$$

With these notations, the BFGS update is written as:

$$H^{n+1}\delta a = H^n \delta a + \frac{(y^n, \delta a)}{(y^n, s^n)} y^n - \frac{(H^n s^n, \delta a)}{(H^n s^n, s^n)} H^n s^n. \tag{A2.21}$$

In the finite dimensional case, if the inner product is the usual Euclidean product, this formula takes the more common form:

$$H^{n+1} = H^n + \frac{y^n (y^n)^T}{(y^n)^T s^n} - \frac{(H^n s^n)(H^n s^n)^T}{(H^n s^n)^T s^n}. \qquad [A2.22]$$

It updates H^n by adding a rank 2 matrix. Its key properties are summarized in Proposition A2.4.

PROPOSITION A2.4.– If H^n is symmetric and positive definite and if $(y^n, s^n) > 0$, then H^{n+1} is symmetric and positive definite and $d^{n+1} = -(H^{n+1})^{-1}\nabla J(a^{n+1})$ is a descent direction.

It can be demonstrated (see [BON 97]) that if Wolfe's linear search is used, the previous condition is verified.

By combining the two previous ideas, we obtain (according to some hypotheses) a globally convergent method. It is possible to state the following result, taken from [BON 97]

THEOREM A2.1.– Suppose that J is convex and of class C^2 in a neighborhood of the set $\{a \in M,\ \ J(a) \leq J(a^0)\}$, and in addition assume that the sequence $J(a^n)$ is bounded from below. Then, the sequence generated by the BFGS algorithm with Wolfe's rule for the line search satisfies:

$$\lim_{n\to\infty} \inf \|\nabla J(a^n)\| = 0.$$

This theorem means that, modulo technical hypotheses, any accumulation point of the computed sequence is a critical point of the functional J.

Regarding the implementation, there is an issue that remains to be solved: how to best use formula [A2.22] without storing the full matrix. In fact, it is possible to provide a formula similar to [A2.22], but on $(H^n)^{-1}$, which makes it possible to directly update d^n, without inverting (explicitly) the linear system. This results in a low-cost per iteration (excepted for the calculation of $J(a)$ and of its gradient).

A2.4. Nonlinear least squares and the Gauss–Newton method

Although the method discussed in the previous section gives complete satisfaction, it applies to general functions, and it is possible to try to make use of the particular structure of problem [6.10]. Effectively, as we have seen in section A2.2, the cost functions that are of interest here have the particular form

$$J(a) = \frac{1}{2} \|\Phi(a) - d\|^2,$$

and we recall how to compute the gradient and the Hessian of this functional, assuming that the Jacobian and the second derivative of the mapping Φ are known.

PROPOSITION A2.5.– The gradient and the Hessian of the functional J are given by the following formulas:

$$\nabla J(a) = \Phi'(a)^*(\Phi(a) - d); \quad \text{[A2.23]}$$

and

$$\nabla^2 J(a)\delta a = \Phi'(a)^*\Phi'(a)\delta a + (\Phi''(a)\delta a)^*(\Phi(a) - d), \text{ for } \delta a \in M. \quad \text{[A2.24]}$$

PROOF.–

– If we let $N(u) = 1/2\,\|u\|^2$ for $u \in D$, then $J(a) = N(\Phi(a) - d)$, and the differentiation of a composed function yields:

$$J'(a)\delta a = N'(\Phi(a) - d))\Phi'(a)\delta a$$

and it is easy to see that $N'(u)\delta u = (u, \delta u)$ for $\delta u \in D$. The result then follows from [A2.4].

– The above equation is differentiated to obtain the expression of the Hessian.

□

Note that if the optimization problem is consistent, that is if the solution $\hat{a}$ satisfies $\Phi(\hat{a}) = d$, or if Φ is linear, then the second term of the Hessian in [A2.24] disappears, and we get:

$$\nabla^2 J(\hat{a}) = \Phi'(\hat{a})^*\Phi'(\hat{a}).$$

This remark suggests that, at least in the neighborhood of the solution, it is possible to neglect the second term of the Hessian. The method obtained by replacing, in Newton's method, the Hessian by the expression

$$H_{\text{GN}}(a) = \Phi'(a)^*\Phi'(a) \quad \text{[A2.25]}$$

is called the *Gauss–Newton method*. An iteration of the method leads to the solution of the linear system

$$\Phi'(a)^*\Phi'(a)d = -\nabla J(a) = -\Phi'(a)^*(\Phi(a) - d), \qquad [A2.26]$$

that is a *linear* least squares problem for the operator $\Phi'(a)$, a problem that has been studied in Appendix 1.

This simplification presents several advantages:

– first, it avoids having to compute the second derivative of Φ, which may be difficult in practice;

– next, it can be seen that the computational cost of the approximated Hessian $H_{\mathrm{GN}}(a)$ is essentially the same as that of the computation of the gradient. Therefore, the cost of a Gauss–Newton iteration is of the same order as that of a descent iteration;

– finally, the approximated Hessian is positive semi-definite, which is not necessarily the case of the full Hessian.

The Gauss–Newton method suffers from the same flaws as the Newton method, in particular the lack of global convergence, and the fact that the method is not defined if $\Phi'(a)^*\Phi'(a)$ is not invertible (positive definite), which may ocurr. A natural idea is then to *regularize* the linearized problems according to the techniques discussed in Chapter 5, in particular Tikhonov's regularization. The corresponding method is called the Levenberg–Marquardt method. This method thus consists of computing $a^{n+1} = a^n + d^n$ where d^n solves the linear system

$$(\varepsilon^2 I + \Phi'(a)^*\Phi'(a))d^n = -\Phi'(a)^*(\Phi(a) - d) \qquad [A2.27]$$

In fact, as we have seen in Chapter 5, it is not necessary to form the operator $\varepsilon^2 I + \Phi'(a)^*\Phi'(a)$, we can solve the least squares problem:

$$\min \frac{1}{2} \left\| \begin{bmatrix} \Phi'(a^n) \\ \varepsilon I \end{bmatrix} d^n + \begin{bmatrix} \Phi(a^n) - d \\ 0 \end{bmatrix} \right\|^2.$$

Naturally, this method can be combined with the Wolfe rule to provide a globally convergent method (under reasonable hypotheses). As a matter of fact, the analysis shows (and the experiment confirms) the method is effective for problems where the final residual $\Phi(\hat{a}) - d$ is small, and may not converge if the final residual is too large.

The choice of the parameter ε is not easier than in Chapter 5. A possibility based on the *trust-region* methods is to adjust the parameter according to the fit of the functional J to a quadratic model (see [KEL 99] for more details).

We can observe that the gradient is an essential ingredient of the BFGS algorithm. For the Gauss–Newton, it seems that the full Jacobian of Φ is required. In Chapter 7, we have developed an efficient computational method for the problems being addressed in this book. We saw that it is possible to compute the gradient of J without computing the Jacobian of Φ (which may seem surprising). This result favors the BFGS method (that we recommend in general), but the adjoint state method of section 6.3.3 enables the computation of a line of the Jacobian of Φ, which can allow the use of the Gauss–Newton method.

Appendix 3

Some Results from Functional Analysis

We recall in this appendix the main functional analysis results needed in this book, as well as additional results concerning operators in Hilbert spaces.

We will give a few proofs, directing the reader to the (numerous) books on the subject. We also particularly cite [BRE 11]. The books [BAU 87, KRE 89, KIR 96], and [GRO 93] also contain a review of functional analysis geared towards application to inverse problems.

For simplicity, we only consider vector spaces over $\mathbf{R}$. In addition, we only consider *separable* spaces, that is to say spaces containing a countable and dense part.

A3.1. Hilbert spaces

We begin by recalling a few definitions.

A3.1.1. *Definitions and examples*

DEFINITION A3.1.– *Let E be a vector space over $\mathbf{R}$. A* norm *on E is a function from E to $\mathbf{R}$, with the following properties:*

– $\forall x \in E,\ \|x\|_E \geq 0$ *and* $\|x\|_E = 0 \Rightarrow x = 0$;

– $\forall x \in E, \forall \alpha \in \mathbf{R},\ \|\alpha\, x\|_E = |\alpha|\, \|x\|_E$;

– $\forall (x, y) \in E^2,\ \|x + y\|_E \leq \|x\|_E + \|y\|_E$.

EXAMPLE A3.1.– In the case where E is of dimension d (we identify it then to $\mathbf{R}^d$), the following norms are the most used:

– $\|x\|_1 = \sum_{i=1}^{d} |x_i|;$

– $\|x\|_2 = \left(\sum_{i=1}^{d} |x_i|^2\right)^{1/2};$

– $\|x\|_\infty = \max_{i=1,\dots,d} |x_i|;$

DEFINITION A3.2.– *Let E be a vector space over $\mathbf{R}$. An inner product over E is a function from $E \times E$ to R, noted $(.,\,.)$, with the following properties:*

– $\forall (x, y, z) \in E^3,\ \forall (\alpha, \beta) \in \mathbf{R}^2,\ (\alpha\, x + \beta\, y,\, z) = \alpha(x,\, z) + \beta(y,\, z)$;

– $\forall (x, y) \in E^2,\ (x,\, y) = (y,\, x)$;

– $\forall x \in E, (x,\, x) \geq 0$;

– $(x,\, x) = 0 \Rightarrow x = 0$.

A vector space equipped with an inner product is called a pre-Hilbert *space.*

EXAMPLE A3.2.– In $\mathbf{R}^d$, the usual Euclidean inner product is as follows:

$$(x,\, y) = \sum_{i=1}^{d} x_i\, y_i. \qquad \text{[A3.1]}$$

EXAMPLE A3.3.– Let Ω be a domain in $\mathbf{R}^d$. The vector space of square integrable functions over Ω:

$$L^2(\Omega) = \{f : \Omega \mapsto \mathbf{R},\ \int_\Omega |f(x)|^2\ dx \leq \infty\} \qquad \text{[A3.2]}$$

is a pre-Hilbert space if it is equipped with the inner product:

$$(f,\, g) = \int_\Omega f(x)\, g(x)\, dx. \qquad \text{[A3.3]}$$

An inner product over E defines a norm on E by means of the following definition:

$$\|x\|_E = \sqrt{(x,\, x)}. \qquad \text{[A3.4]}$$

Among the three norms of example A3.1, only the second originates from an inner product (that of example A3.2).

DEFINITION A3.3.– *A* Hilbert space *is a vector space equipped with an inner product that is complete for the norm associated with this inner product.*

EXAMPLE A3.4.– The vector space $\mathbf{R}^d$, equipped with the usual Euclidean inner product, is a Hilbert space.

The following result is paramount.

PROPOSITION A3.1.– Let Ω be an open domain of $\mathbf{R}^d$. The vector space $L^2(\Omega)$, equipped with the inner product defined by [A3.3], is a Hilbert space.

EXAMPLE A3.5 (Sobolev Space).– To simplify, we consider one dimension, over the interval $[0,1]$. The first-order Sobolev space is given as:

$$H^1(0,1) = \left\{ u \in L^2(0,1), \exists u_1 \in L^2(0,1) \text{ such that} \right.$$

$$\left. \forall \varphi \in C_c^1(0,1), \int_0^1 u(t)\,\varphi'(t)\,dt = -\int_0^1 u_1(t)\varphi(t)\,dt \right\}, \quad \text{[A3.5]}$$

(where $C_c^1(0,1)$ is the space of continuously differentiable functions, with compact support in $[0,1]$). This definition is equivalent to that, more commonly used in distributions theory. For $u \in H^1(0,1)$, we denote $u' = u_1$.

It can be shown (see [BRE 11]) that $H^1(0,1)$ is a Hilbert space if it is equipped with the inner product:

$$(u,v)_{H^1} = \int_0^1 u(t)\,v(t)\,dt + \int_0^1 u'(t)\,v'(t)\,dt. \quad \text{[A3.6]}$$

In applications, there is often a need for the subspace of H^1 corresponding to the functions equal to zero on the boundary (in the sense of the trace). This subspace is denoted by $H_0^1(0,1)$ and can be equipped with the following inner product:

$$(u,v)_{H_0^1} = \int_0^1 u'(t)\,v'(t)\,dt. \quad \text{[A3.7]}$$

It is shown (this is a consequence of the Poincaré inequality, see again [BRE 11]) that the corresponding norm is equivalent to the norm induced by that of the space H^1.

A3.1.2. *Properties of Hilbert spaces*

PROPOSITION A3.2 (Cauchy–Schwarz inequality).– For all $(x, y) \in E^2$, we have the inequality:

$$|(x,\, y)|^2 \le \|x\|_E^2\, \|y\|_E^2 \,. \qquad \text{[A3.8]}$$

Equality holds only if x and y are proportional.

PROPOSITION A3.3 (Parallelogram identity).– For all $(x, y) \in E^2$, we have the identity:

$$\|x+y\|_E^2 + \|x-y\|_E^2 = 2\left(\|x\|_E^2 + \|y\|_E^2\right). \qquad \text{[A3.9]}$$

The following result is one of the most important of the theory.

THEOREM A3.1 (Projection theorem).– Let K be a closed, convex subset of E, and let $z \in E$ be given. There is a unique element $x_0 \in K$ such that:

$$\|z - x_0\|_E = \inf \|y - x\|_E \,,\ x \in K. \qquad \text{[A3.10]}$$

The point x_0 is characterized by the condition:

$$x_0 \in K \ \text{ and } \ (z - x_0,\, x - x_0) \le 0,\ \forall x \in K. \qquad \text{[A3.11]}$$

The point x_0 exhibited in theorem A3.1 is called the *projection* of z over K. In the case in which K is a vector subspace (then denoted by F), this result can be made more precise:

COROLLARY A3.1.– Let F a closed subspace of E and let $z \in E$. The projection of z over F is characterized by:

$$x_0 \in F \ \text{ and } \ (z - x_0,\, x) = 0,\ \forall x \in F. \qquad \text{[A3.12]}$$

In a Hilbert space, two vectors are said to be *orthogonal* if their inner product is zero. The orthogonal of a vector subspace is:

$$F^\perp = \{x \in E,\ (x,\, y) = 0,\ \forall y \in F\}\,.$$

COROLLARY A3.2.– Let F be a vector subspace of a Hilbert space E (not necessarily closed). Then,

$$F^{\perp} + \overline{F} = E. \tag{A3.13}$$

A3.1.3. *Hilbert bases*

DEFINITION A3.4.– *A Hilbert basis of a Hilbert space E is a sequence* $(e_n)_{n\in\mathbf{N}^*}$ *such that:*

– $\|e_n\|_E = 1$, $\forall n$, *and* $(e_n, e_m) = 0$, $\forall n \neq m$;

– *the vector space spanned by the* (e_n) *is dense in* E.

We specify the second condition: let $F_n = \operatorname{span}\{e_1, \ldots, e_n\}$. The subspaces F_n are nested ($F_n \subset F_m$ for $n \leq m$), so that $F = \cup_{n\in N^*} F_n$ is a vector subspace. The second condition of the definition expresses that this subspace is dense in E, that is any element of E can be arbitrarily approximated by an element of F.

It can be shown that every (separable) vector space admits a Hilbert basis. Given a Hilbert basis $(e_n)_{n\in\mathbf{N}}$ of E, any element of E is written as:

$$x = \sum_{n=1}^{\infty} (x, e_n)\, e_n, \tag{A3.14}$$

with (this is the Bessel–Parseval equality):

$$\|x\|_E^2 = \sum_{n=1}^{\infty} |(x, e_n)|^2 . \tag{A3.15}$$

Such an expansion is unique, that is if there is an expansion

$$x = \sum_{n=1}^{\infty} x_n e_n$$

with $\sum_{n=1}^{\infty} |x_n|^2 < \infty$, then $x_n = (x,\, e_n)$. Note, however, that a Hilbert basis *is not* an algebraic basis, since this expansion is not a finite linear combination.

Obviously, an orthogonal basis of a finite dimension vector space is a Hilbert basis. We know how to explicitly build Hilbert bases for certain L^2 spaces.

EXAMPLE A3.6.– The two sequences of functions

$$\left(\sqrt{\frac{2}{\pi}}\sin nx\right)_{n\geq 1}, \text{ or } \left(\sqrt{\frac{2}{\pi}}\cos nx\right)_{n\geq 0}$$

are Hilbert bases of $L^2(0,\pi)$. In this case, the expansion obtained in [A3.14] is identified with a Fourier series expansion (after extension by imparity and periodicity).

A3.2. Linear operators in Hilbert spaces

Functional analysis makes topology and linear algebra interact. Thus, on a Hilbert space, it will be natural to study the mappings that are consistent with both the structure of the vector space (linear applications) and the Hilbertian structure (the continuous applications).

A3.2.1. *General properties*

DEFINITION A3.5.– *An operator (linear, continuous) A from a Hilbert space E to a Hilbert space F is a continuous linear mapping from E to F, that satisfies:*

– $\forall u \in E, Au \in F$;

– $\forall (u,v) \in E \times E, \forall (\alpha,\beta) \in \mathbf{R}^2,\ A(\alpha u + \beta v) = \alpha Au + \beta Av$;

– $\exists M > 0, \forall u \in E, \|Au\|_F \leq M\|u\|_E$.

The smallest number M that satisfies the third point above is called the norm *of the operator A:*

$$\|A\| = \sup_{u\in E}\frac{\|Au\|_F}{\|u\|_E}. \qquad \text{[A3.16]}$$

This is the same concept as the matrix norm used in linear algebra (see, for example, [DEM 97, GOL 96]).

We recall the two fundamental spaces associated with a linear operator:

– The kernel of A is the subspace of E: $\operatorname{Ker} A = \{u \in E,\ Au = 0\}$;

– The image of A is the subspace of F: $\operatorname{Im} A = \{v \in F, \exists u \in E,\ Au = v\}$.

Note that although $\mathrm{Ker}\, A$ is always closed, as a reciprocal image of the closed subspace 0 of F, $\mathrm{Im}\, A$ may not be closed (see the discussion about this point in theorem 4.1).

An operator is said to be of *finite rank* if and only if its image is a vector subspace of finite dimension of F.

The following theorem is one of the fundamental results of the theory of linear operators.

THEOREM A3.2 (Open mapping theorem).– Let A be a linear operator from E to F. The image by A of an open subset of E is an open subset of F.

In particular, the inverse of a bijective and continuous linear operator is continuous.

In the case where the image space F is the field of scalars, the operator is called a *linear form*. The vector space of linear continuous forms is called the dual space of E, and is denoted by E'. In the case of a Hilbert space, the dual can be identified in a canonical manner with the space itself.

THEOREM A3.3 (Riesz's).– Let L a continuous linear form over E. There is a unique vector $x_0 \in E$ such that

$$L(x) = (x_0, x), \ \forall x \in E. \qquad [A3.17]$$

A3.2.2. *Adjoint operator*

THEOREM A3.4.– Let A be a continuous linear operator from E to F. There is a unique operator from F to E, denoted byA^*, such that:

$$\forall u \in E, \forall v \in F, \ (Au, v) = (u, A^* v). \qquad [A3.18]$$

This operator is called the *adjoint* of A. In addition, it satisfies: $(A^*)^* = A$ and $\|A^*\| = \|A\|$.

PROOF.– First, let $v \in F$ be fixed. The mapping

$$E \mapsto \mathbf{R}$$

$$u \to (Au, v)$$

is linear and continuous, since $|(Au, v)| \leq \|A\| \, \|u\|_E \, \|v\|_F$. According to the Riesz theorem, there is a unique element of E, denoted for the time by v^*, such that

$$(Au, v) = (u, v^*), \ \forall u \in E.$$

It still has to be proved that the mapping $v \to v^*$ is well defined, linear and continuous. It will be then possible to write $A^*v = v^*$, and A^* will in fact have the mentioned properties.

The fact that our mapping is well-defined results from the uniqueness in the Riesz theorem. Moreover, the same also holds for the linearity (using the linearity of the dot product). We can note that the definition is symmetric in A and A^*, so that $(A^*)^* = A$.

Finally, continuity is a consequence of the Cauchy–Schwarz inequality:

$$\|A^*v\|_E^2 = (A^*v,\, A^*v) = (AA^*v,\, v) \leq \|A\|\; \|A^*v\|_E\; \|v\|_F\,,$$

that shows in addition that $\|A^*\| \leq \|A\|$. Since we have just seen that A and A^* act in a similar way, we finally get $\|A^*\| = \|A\|$. □

REMARK A3.1.– In the finite dimensional case, identifying the linear operator A with its matrix in orthogonal bases of $\mathbf{R}^n$ and $\mathbf{R}^p$, it can be seen that the matrix of the adjoint operator is none other than the transposed matrix of A (define $u = e_i$, $i = 1, \ldots, n$ and $v = f_j$, $j = 1, \ldots, p$ elements of the bases under consideration).

The following proposition gathers some simple properties of the adjoint.

PROPOSITION A3.4.– Let A and B be two linear operators, α and β two scalars.

– Linearity: $(\alpha A + \beta B)^* = \alpha A^* + \beta B^*$;

– Composition: $(AB)^* = B^* A^*$.

There are notable relations between the kernel and the image of an operator and those of its adjoint.

PROPOSITION A3.5.– We have the following relations (where $\overline{X}$ indicates the closure of the set X) :

– $\operatorname{Ker} A^* = (\operatorname{Im} A)^{\perp}$;

– $(\operatorname{Ker} A)^{\perp} = \overline{\operatorname{Im} A^*}$.

DEFINITION A3.6.– *An operator in E is called self-adjoint if and only if:*

$$\forall (u, v) \in E \times E,\; (Au, v) = (u, Av) \tag{A3.19}$$

REMARK A3.2.– In the finite dimensional case, the self-adjoint operators are those with a symmetric matrix.

A3.2.3. *Compact operators*

DEFINITION A3.7.– *A linear operator A from E to F is called* compact *if and only if the image of any bounded subset of E is relatively compact in F.*

REMARK A3.3.– This condition means that if $B \subset E$ is bounded, $\overline{A(B)}$ (the closure of $A(B)$) is compact in F.

EXAMPLE A3.7.–

– any finite rank operator is compact;

– the canonical injection of $H_0^1(0,1) \to L^2(0,1)$ is compact (more generally, $]0,1[$ can be replaced by a regular open bounded domain of $\mathbf{R}^d$);

– an important class of compact operators is provided by integral operators, studied in Chapter 3.

We start with some basic properties of compact operators.

PROPOSITION A3.6.– Let E, F, G be three Hilbert spaces:

1) The set of the compact operators from E to F is a vector subspace of $\mathcal{L}(E,F)$.

2) If $A_1 \in \mathcal{L}(E,F)$ is compact and $A_2 \in \mathcal{L}(F,G)$, then $A_2A_1 \in \mathcal{L}(E,G)$ is compact.

3) If $A_1 \in \mathcal{L}(E,F)$ and $A_2 \in \mathcal{L}(F,G)$ is compact, then $A_2A_1 \in \mathcal{L}(E,G)$ is compact.

4) If $A \in \mathcal{L}(E,F)$ is compact, $A^* \in \mathcal{L}(F,E)$ is also compact.

5) Let $(A_n)_{n\in\mathbf{N}}$ be a sequence of compact operators of E in F. If A_n converges to A in $\mathcal{L}(E,f)$, that is to say, if

$$\|A_n - A\| = \sup_{u\neq 0} \frac{\|A_n u - Au\|_F}{\|u\|_E} \underset{n\to\infty}{\longrightarrow} 0,$$

then A is compact.

The following theorem provides (in the case of Hilbert spaces) a characterization that is both useful and closer to intuition.

THEOREM A3.5.– An operator of E in F is compact if and only if it is the limit of a sequence of finite rank operators.

This theorem means that compact operators are those that most "resemble" the usual finite dimensional operators. It should be noted that this result is no longer valid if E and F are Banach spaces.

However, there is a difference between the finite dimensional and infinite dimensional cases that will be essential to the study of ill-posed problems.

PROPOSITION A3.7.– If E is *not* finite dimensional, then the identity $E \to E$ is never compact.

REMARK A3.4.– We do not provide the proof (see, for example, [BRE 11]), which is a consequence of the fact that the unit sphere of a normed vector space of finite dimension is not compact.

COROLLARY A3.3.– Let A be a compact operator from E to F, where E and F are two Hilbert spaces that are not of finite dimension. Then, A is never invertible in $\mathcal{L}(E,F)$.

PROOF.– If A were invertible, its inverse A^{-1} would satisfy:

$$AA^{-1} = I.$$

Since A is compact and since the identity cannot be compact, according to proposition A3.7, we have a contradiction. □

REMARK A3.5.– In the previous corollary, the (algebraic) inverse of A may or may not exist (A may or may not be injective), but if it exists, it will not be continuous. This is related to the unclosed character of the image of A and this is discussed in Chapter 5.

To conclude, we quote without proof an "abstract" version of the Fredholm alternative. This result concerns the equations of the type

$$(I - A)u = f \qquad \text{[A3.20]}$$

in which A is a compact operator in E.

THEOREM A3.6.– Let A be a compact operator in a Hilbert space E:

– the kernel $\mathrm{Ker}\,(I - A)$ is finite dimensional and the image $\mathrm{Im}\,(I - A)$ is closed in E;

– if $I - A$ is injective, it is also surjective, and then the inverse $(I - A)^{-1}$ is continuous. If $I - A$ is not injective, equation [A3.20] has a solution if and only if $f \in \mathrm{Ker}\,(I - A)^{\perp}$.

REMARK A3.6.–

– this theorem extends to operators of the form I+compact the similar well-known result in finite dimension (for any linear equations systems);

– the second point of the theorem means that equation [A3.20] only has a solution if the right-hand side meets the orthogonality conditions of the kernel (cf finite dimension).

A3.3. Spectral decomposition of compact self-adjoint operators

In this section, A denotes a compact self-adjoint operator in E.

DEFINITION A3.8.– *The* spectrum *of A is the set*

$$\sigma(A) = \{\lambda \in \mathbf{C},\ A - \lambda I \textit{ is not invertible in } \mathcal{L}(E)\}.$$

DEFINITION A3.9.– *A number $\lambda \in \mathbf{C}$ is an* eigenvalue *if and only if $A - \lambda I$ is not injective.*

It should be remembered that, *a priori*, if $\lambda \in \sigma(A)$, three cases can occur for the operator $A - \lambda I$:

– it may not be injective, and in this case λ is an eigenvalue;

– it may not be surjective;

– it may be bijective, but the inverse is not continuous.

It can be shown that in fact, if A is compact and $\lambda \neq 0$, the two last alternatives cannot occur.

To simplify the statement of the following theorem, we will make the assumption that A is not of finite rank (if this were the case, there would be a finite number of eigenvalues and 0 would not be an eigenvalue).

PROPOSITION A3.8.– Denote by Λ the set of the eigenvalues of A.

1) $\sigma(A) = \{0\} \cup \Lambda$;

2) any nonzero eigenvalue is of finite multiplicity;

3) A has at most a countably infinite number of eigenvalues, whose only possible accumulation point is 0;

4) the eigenvalues of A are real and eigenvectors corresponding to distinct eigenvalues are orthogonal;

5) one of the numbers $\pm \|A\|$ is an eigenvalue of A.

The nonzero eigenvalues of a compact self-adjoint operator can therefore be ordered in a sequence that tends toward 0.

We can now state the most important result of this section.

THEOREM A3.7.– Denote by Λ the set of eigenvalues of A, with $\lim_{n\to\infty} \lambda_n = 0$. There is a Hilbert basis $(e_n)_{n\in\mathbf{N}}$ of $(\text{Ker } A)^\perp$ such that

$$\forall x \in E,\ x = x_0 + \sum_{n\in\mathbf{N}} (x, e_n)e_n,\ Ax = \sum_{n\in\mathbf{N}} \lambda_n (x, e_n)e_n, \qquad \text{[A3.21]}$$

where $x_0 \in \text{Ker } A$.

Naturally, the series in [A3.21] are to be considered in the sense of the norm of E. Here again, if A is of finite rank, the sums above are in fact finite.

It is clear that any operator given by a formula such as [A3.21] is compact. An "operating" definition of the compact operators is an operator given by equation [A3.21], where the sequence $(\lambda_n)_{n\in\mathbf{N}})$ is a real sequence that tends toward 0.

This result is the analog of the diagonalization of symmetric matrices in an orthonormal basis. In the Hilbert basis given by the theorem, the action of A is diagonal.

Bibliography

[ADA 13] ADAMS B., BAUMAN L., BOHNHOFF W. *et al.*, A multilevel parallel object-oriented framework for design optimization, parameter estimation, uncertainty quantification, and sensitivity analysis: version 5.4 user's manual, Report no. SAND2010-2183, Sandia Technical Report, 2013.

[AKI 80] AKI K., RICHARDS P., *Quantitative Seismology: Theory and Methods*, Freeman, San Francisco, 1980.

[ARN 12] ARNOLD V.I., *Équations différentielles ordinaires*, 3rd ed., Collection MIR, Ellipses, Paris, 2012.

[AST 05] ASTER R.C., BORCHERS B., THURBER C.H., *Parameter Estimation and Inverse Problems*, 2nd ed., Elsevier Academic Press, Boston, 2005.

[BAM 77] BAMBERGER A., CHAVENT G., LAILLY P., "Une application de la théorie du contrôle à un problème inverse de sismique", *Annales de géophysique*, vol. 33, pp. 183–199, 1977.

[BAM 79] BAMBERGER A., CHAVENT G., LAILLY P., "About the stability of the inverse problem in 1D wave equations – applications to the interpretation of seismic profiles", *Applied Mathematics and Optimization*, vol. 5, pp. 1–47, 1979.

[BAM 82] BAMBERGER A., CHAVENT G., HEMON C. *et al.*, "Inversion of normal incidence seismograms", *Geophysics*, vol. 47, no. 5, pp. 757–770, 1982.

[BAN 89] BANKS H.T., KUNISCH K., *Estimation Techniques for Distributed Parameter Systems*, Birkhäuser-Verlag, Zürich, 1989.

[BAU 87] BAUMEISTER J., *Stable Solution of Inverse Problems*, Vieweg, Braunschweig, 1987.

[BEA 87] BEAR J., VERRUIJT A., *Modeling Groundwater Flow and Pollution*, Springer, Amsterdam, vol. 2, 1987.

[BJÖ 96] BJÖRCK N, *Numerical Methods for Least Squares Problems*, SIAM, Philadelphia, 1996.

[BLE 00] BLEISTEIN N., COHEN J.K., STOCKWELL JR J.W., "Mathematics of multidimensional seismic imaging, migration and inversion", *Interdisciplinary Applied Mathematics*, Springer, New York, vol. 13, 2000.

[BON 97] BONNANS J.F., GILBERT J., LEMARÉCHAL C., *et al.*, *Optimisation numérique – Aspects théoriques et pratiques*, *Mathématiques et Applications*, Springer, Berlin, vol. 27, 1997.

[BRE 11] BREZIS H., *Functional Analysis, Sobolev Spaces and Partial Differential Equations*, Universitext, Springer, New York, 2011.

[BUI 13] BUI-THANH T., GHATTAS O., MARTIN J. *et al.*, "A computational framework for infinite-dimensional Bayesian inverse problems Part I: The linearized case, with application to global seismic inversion", *SIAM Journal on Scientific Computing*, vol. 35, no. 6, pp. A2494–A2523, 2013.

[BUI 15] BUI-THANH T., GHATTAS O., "A scalable algorithm for MAP estimators in Bayesian inverse problems with Besov priors", *Inverse Problem and Imaging*, vol. 9, no. 1, pp. 27–53, 2015.

[CAL 07] CALVETTI D., SOMERSALO E., *An Introduction to Bayesian Scientific Computing. Ten Lectures in Subjective Scientific Computing*, vol. 2, Springer, New York, 2007.

[CAN 84] CANNON J.R., *The One-dimensional Heat Equation*, Addison-Wesley, MA, 1984.

[CAO 03] CAO Y., LI S., PETZOLD L. *et al.*, "Adjoint sensitivity analysis for differential-algebraic equations: the adjoint DAE system and its numerical solution", *SIAM Journal on Scientific Computing*, vol. 24, no. 3, pp. 1076–1089, 2003.

[CHA 79] CHAVENT G., "Identification of distributed parameter systems: about the output least squares method, its implementation and identifiability", *Proceedings of the 5th IFAC Symposium on Identification and System Parameter Estimation*, Pergamon Press, Oxford, pp. 85–97, 1979.

[CHA 87] CHAVENT G., "Identifiability of parameters in the output least squares formulation", in WALTER E. (ed.), *Structural Identifiability of Parametric Models*, Pergamon Press, Oxford, 1987.

[CHA 91] CHAVENT G., "On the theory and practice of non-linear least squares", *Advances in Water Resources*, vol. 14, no. 2, pp. 55–63, 1991.

[CHA 96] CHAVENT G., Problèmes inverses, notes de cours de DEA, Paris Dauphine University, 1996.

[CHA 02] CHAVENT G., KUNISCH K., "The output least squares identifiability of the diffusion coefficient from an H^1-observation in a 2-D elliptic equation", *ESAIM: Control, Optimisation and Calculus of Variations*, vol. 8, pp. 423–440 , 2002.

[CHA 09] CHAVENT G., *Nonlinear Least Squares for Inverse Problems*, Scientific Computation, Springer, New York, 2009.

[CHO 09] CHOULLI M., "Une introduction aux problèmes inverses elliptiques et paraboliques", *Mathématiques & Applications*, Springer-Verlag, Berlin, vol. 65, 2009.

[CIA 82] CIARLET P.G., *Introduction à l'analyse numérique matricielle et à l'optimisation*, Dunod, Paris, 1982.

[CLÉ 01] CLÉMENT F., CHAVENT G., GÒMEZ S., "Migration-based traveltime waveform inversion of 2-D simple structures: a synthetic example", *Geophysics*, vol. 66, no. 3, pp. 845–860, 2001.

[COL 92] COLTON D., KRESS R., *Inverse Acoustic and Electromagnetic Scattering Theory*, Springer, New York, 1992.

[COL 00] COLEMAN T.F., SANTOSA F., VERMA A., "Efficient calculation of Jacobian and adjoint vector products in wave propagational inverse problem using automatic differentiation", *Journal of Computational Physics*, vol. 157, pp. 234–255, 2000.

[DAS 16] DASHTI M., STUART A.M., "The bayesian approach to inverse problems", in in GHANEM R., HIGDON D., OWHADI H. (eds), *Handbook of Uncertainty Quantification*, Springer, Berlin, 2016.

[DAU 90] DAUTRAY R., LIONS J.L. (eds), *Mathematical Analysis and Numerical Methods for Science and Technology*, Springer, Berlin-Heidelberg, 1990.

[DEM 97] DEMMEL J.W., *Applied Numerical Linear Algebra*, SIAM, Philadelphia, 1997.

[DEN 96] DENNIS J.E., SCHNABEL R.B., *Numerical Methods for Unconstrained Optimization and Nonlinear Equations*, SIAM, Philadelphia, 1996.

[DOL 04] DOLAN E.D., MORÉ J.J., MUNSON T.S., Benchmarking Optimization Software with COPS 3.0, Technical Report no. ANL/MCS-273, Mathematics and Computer Science Division, Argonne National Laboratory, 2004.

[DRI 97] DRISCOLL T.A., "Eigenmodes of isospectral drums", *SIAM Review*, vol. 39, no. 1, pp. 1–17, 1997.

[DUC 13] DUCHATEAU P., "An adjoint method for proving identifiability of coefficients in parabolic equations", *Journal of Inverse and Ill-Posed Problems*, vol. 21, no. 5, pp. 639–663, 2013.

[ENG 93] ENGL H.W., "Regularization methods for the stable solution of inverse problems", *Surveys of Mathematics for Industry*, vol. 3, pp. 71–143, 1993.

[ENG 96] ENGL H.W., HANKE M., NEUBAUER A., *Regularization of Inverse Problems*, Kluwer, Dordrecht, 1996.

[EPA 08] EPANOMERITAKIS I., AKÇELIK V., GHATTAS O. *et al.*, "A Newton-CG method for large-scale three-dimensional elastic full-waveform seismic inversion", *Inverse Problems*, vol. 24, no. 3, pp. 034–015, 2008.

[ERN 13] ERN A., GUERMOND J.-L., *Theory and Practice of Finite Elements*, Springer, 2013.

[FAR 88] FAREBROTHER R.W., *Linear Least Squares Computations*, Marcel Dekker Inc., New York, 1988.

[GIL 91] GILBERT J.C., LE VEY G., MASSE J., La Différentiation automatique de fonctions représentées par des programmes, Report no. RR-1557, INRIA, 1991.

[GOC 99] GOCKENBACH M.S., PETRO M.J., SYMES W.W., "C++ classes for linking optimization with complex simulations", *ACM Transactions on Mathematical Software*, vol. 25, no. 2, pp. 191–212, 1999.

[GOC 01] GOCKENBACH M.S., REYNOLDS D.R., SYMES W.W., "Automatic differentiation and the adjoint state method", in CORLISS G., FAURE C., GRIEWANK A. *et al.* (eds), *Automatic Differentiation of Algorithms: From Simulation to Optimization*, Springer, New York, 2001.

[GOC 02] GOCKENBACH M.S., REYNOLDS D.R., SHEN P., *et al.*, "Efficient and automatic implementation of the adjoint state method", *ACM Transactions on Mathematical Software (TOMS)*, vol. 28, no. 1, pp. 22–44, 2002.

[GOC 06] GOCKENBACH M.S., *Understanding and Implementing the Finite Element Method*, Siam, Philadelphia, PA, 2006.

[GOL 96] GOLUB G.H., LOAN C. F.V., *Matrix Computations*, 3rd ed., Johns Hopkins University Press, Baltimore, 1996.

[GOR 91] GORENFLO R., VESSELLA S., *Abel Integral Equations. Analysis and Applications*, Springer, Berlin, 1991.

[GRI 93] GRIEWANK A., Some Bounds on the Complexity of Gradients, Jacobians, and Hessians, Report no. MCS-P355-0393, Mathematics and Computer Science Division, Argonne National Laboratory, Argonne, IL, 1993.

[GRI 00a] GRIEWANK A., *Evaluating Derivatives, Principles and Techniques of Algorithmic Differentiation*, SIAM, Philadelphia, 2000.

[GRI 00b] GRIEWANK A., WALTHER A., "Algorithm 799: revolve: an implementation of checkpointing for the reverse or adjoint mode of computational differentiation", *ACM Trans. Math. Softw.*, vol. 26, no. 1, pp. 19–45, March 2000.

[GRO 93] GROETSCH C.W., *Inverse Problems in the Mathematical Sciences*, Vieweg, Wiesbaden, 1993.

[HAD 23] HADAMARD J., *Lectures on Cauchy's Problem in Linear Partial Differential Equations*, Yale University Press, Yale, 1923.

[HAI 87] HAIRER E., NORSETT S., WANNER G., *Solving Ordinary Differential Equations: 1: Nonstiff Problems*, Springer, Berlin, 1987.

[HAI 91] HAIRER E., WANNER G., *Solving Ordinary Differential Equations: 2: Stiff and Differential-Algebraic Problems*, Berlin, 1991.

[HAN 90] HANSEN P.C., "The discrete Picard condition for discrete ill-posed problems", *BIT*, vol. 30, pp. 658–672, 1990.

[HAN 92] HANSEN P.C., "Analysis of discrete ill-posed problems by means of the L-curve", *SIAM Review*, vol. 34, pp. 561–50, 1992.

[HAN 93a] HANKE M., HANSEN P.C., "Regularization methods for large-scale problems", *Surveys of Mathematics for Industry*, vol. 3, pp. 253–315, 1993.

[HAN 93b] HANSEN P.C., O'LEARY D.P., "The use of the L-curve in the regularization of discrete ill-posed problems", *SIAM Journal on Scientific Computing*, vol. 14, pp. 1487-1503, 1993.

[HAN 95] HANKE M., *Conjugate Gradient Type Methods for Ill-Posed Problems*, Longman, Harlow, 1995.

[HAN 98] HANSEN P.C., *Rank-Deficient and Discrete Ill-Posed Problems: Numerical Aspects of Linear Inversion*, SIAM, Philadelphia, 1998.

[HAN 06] HANSEN P.C., NAGY J.G., O'LEARY D.P., *Deburring Images: Matrices, Spectra, and Filtering*, SIAM, Philadelphia, 2006.

[HAN 07] HANSEN P.C., "Regularizarion tools version 4.0 for MATLAB 7.3,", *Numerical Algorithms*, vol. 46, pp. 189–194, 2007.

[HAN 10] HANSEN P.C., *Discrete inverse problems: Insight and Algorithms*, of *Fundamentals of Algorithms*, SIAM, Philadelphia, vol. 7, 2010.

[HER 80] HERMAN G.T., ed., *Image Reconstruction from Projections: the Fundamentals of Computerized Tomography*, Academic Press, New York, 1980.

[IDI 08] IDIER J., ed., *Bayesian Approach to Inverse Problems*, ISTE Ltd, London et John Wiley & Sons Inc, Hoboken, 2008.

[ISA 98] ISAKOV V., *Inverse Problems for Partial Differential Equations*, Springer, New York, 1998.

[KAC 66] KAC M., "Can one hear the shape of a drum?", *American Mathematical Monthly*, vol. 73, pp. 1–23, 1966.

[KAI 05] KAIPIO J., SOMERSALO E., *Statistical And Computational Inverse Problems*, Springer, 2005.

[KEL 76] KELLER J.B., "Inverse problems", *Amer. Math. Monthly*, vol. 83, pp. 107–118, 1976.

[KEL 99] KELLEY C.T., *Iterative Methods for Optimization*, SIAM, Philadelphia, 1999.

[KER 04] KERN M., "Using Scilab to solve inverse problems for ordinary differential equations", *Scilab Conference*, 2004.

[KIR 96] KIRSCH A., *An Introduction to the Mathematical Theory of Inverse Problems*, Springer, New York, 1996.

[KNA 03] KNABNER P., ANGERMANN L., *Numerical Methods for Elliptic and Parabolic Partial Differential Equations*, Springer Verlag, New York, 2003.

[KRE 89] KRESS R., *Linear Integral Equations*, Springer, Berlin, 1989.

[KUN 88] KUNISCH K., "Inherent identifiability of parameters in elliptic differential equations", *Journal of Mathematical Analysis and Applications*, vol. 132, no. 2, pp. 453–472, 1988.

[LAA 87] VAN LAARHOVEN P., AARTS E., *Simulated Annealing, Theory and Practice*, Kluwer, Dordrecht, 1987.

[LAN 51] LANDWEBER L., "An iteration formula for Fredholm integral equations of the first kind", *American Journal of Mathematics*, vol. 73, pp. 615–624, 1951.

[LAW 95] LAWSON C.L., HANSON R.J., *Solving Least Squares Problems*, SIAM, Philadelphia, 1995.

[LIU 93] LIU J., "A multiresolution method for distributed parameter estimation", *SIAM Journal on Scientific Computing*, vol. 14, pp. 389–405, 193.

[LOU 92] LOUIS A.K., "Medical imaging, state of the art and future developments", *Inverse Problems*, vol. 59, pp. 277–294, 1992.

[MAR 86] DE MARSILY G., *Quantitative Hydrogeology: Groundwater Hydrology for Engineers*, Academic Press, Orlando, FL, 1986.

[MAR 08] MARCHAND E., CLÉMENT F., ROBERTS J.E. *et al.*, "Deterministic sensitivity analysis for a model for flow in porous media", *Advances in Water Resources*, vol. 31, no. 8, pp. 1025-1037, 2008.

[MAR 12] MARTIN J., WILCOX L.C., BURSTEDDE C. *et al.*, "A stochastic Newton MCMC method for large-scale statistical inverse problems with application to seismic inversion", *SIAM Journal on Scientific Computing*, vol. 34, no. 3, pp. A1460–A1487, 2012.

[MÉT 13] MÉTIVIER L., BROSSIER R., VIRIEUX J., OPERTO S., "Full waveform inversion and the truncated Newton method", *SIAM Journal on Scientific Computing*, vol. 35, no. 2, pp. B401–B437, 2013.

[MON 15] MONTEILLER V., CHEVROT S., KOMATITSCH D. *et al.*, "Three-dimensional full waveform inversion of short-period teleseismic wavefields based upon the SEM-DSM hybrid method", *Geophysical Journal International*, 2015.

[MOR 84] MOROZOV V.A., *Methods for Solving Incorrectly Posed problems*, Springer-Verlag, New York, 1984.

[MOR 93] MORÉ J.J., WRIGHT S.J. (eds), *Optimization Software Guide*, SIAM, Philadelphia, 1993.

[MOS 98] MOSÉ R., Habilitation à diriger des recherches, University Louis Pasteur, Strasbourg, 1998.

[MUÉ 12] MUELLER J.L., SILTANEN S., *Linear and Nonlinear Inverse Problems with Practical Applications*, SIAM, Philadelphia, 2012.

[NAT 86] NATTERER F., *The Mathematics of Computerized Tomography*, Wiley, New York, 1986.

[NOC 99] NOCEDAL J., WRIGHT S.J., *Numerical Optimization*, Springer, New York, 1999.

[PAI 82] PAIGE C.C., SAUNDERS M.A., "LSQR: an algorithm for sparse linear equations and sparse least squares", *ACM Transactions on Mathematical Software*, vol. 8, no. 1, pp. 43–71, 1982.

[PET 14] PETRA N., MARTIN J., STADLER G. *et al.*, "A computational framework for infinite-dimensional Bayesian inverse problems, Part II: Stochastic Newton MCMC with application to ice sheet flow inverse problems", *SIAM Journal on Scientific Computing*, vol. 36, no. 4, pp. A1525–A1555, 2014.

[QUA 07] QUARTERONI A., SACCO R., SALERI F., *Méthodes numériques: algorithmes, analyse et applications*, Springer, Milan, 2007.

[RAM 05] RAMM A.G., *Inverse Problems*, Springer, New York, 2005.

[RES 98] RESTREPO J.M., LEAF G.K., GRIEWANK A., "Circumventing storage limitations in variational data assimilation studies", *SIAM Journal on Scientific Computing*, vol. 19, no. 5, pp. 1586–1605, 1998.

[RUD 92] RUDIN L.I., OSHER S., FATEMI E., "Nonlinear total variation based noise removal algorithms", *Physica D: Nonlinear Phenomena*, vol. 60, nos. 1–4, pp. 259–268, 1992.

[SAI 12] SAIBABA A.K., KITANIDIS P.K., "Efficient methods for large-scale linear inversion using a geostatistical approach", *Water Resources Research*, vol. 48, 2012.

[SAL 08] SALTELLI A., RATTO M., ANDRES T. *et al.*, *Global Sensitivity Analysis: The Primer*, John Wiley & Sons, Chichester, 2008.

[SAP 90] SAPORTA G., *Probabilités, analyse des données et statistique*, Technip, 1990.

[SCA 97] SCALES J.A., SNIEDER R., "To Bayes or not to Bayes?", *Geophysics*, vol. 63, pp. 1045–1046, 1997.

[SCA 01] SCALES J.A., TENORIO L., "Prior information and uncertainty in inverse problems", *Geophysics*, vol. 66, no. 2, pp. 389–397, 2001.

[SCH 12] SCHWEDE R.L., NGO A., BASTIAN P. *et al.*, "Efficient parallelization of geostatistical inversion using the quasi-linear approach", *Computers & Geosciences*, vol. 44, pp. 78-85, 2012.

[SEA 71] SEARLE S.R., *Linear Models*, John Wiley & Sons, New York, 1971.

[SEB 77] SEBER G.A.F., *Linear Regression Analysis*, John Wiley & Sons, New York, 1977.

[SER 05] SERBAN R., HINDMARSH A.C., "CVODES, the sensitivity–enabled ode solver in SUNDIALS", *Proceedings of the 5th International Conference on Multibody SYstems, Nonlinear Dynamics and Control*, Long Beach, CA, ASME, 2005.

[SIE 95] SIEGEL P., Transfert de masse en milieux poreux complexes: modélisation et estimation de paramètres par éléments finis mixtes hybrides, PhD Thesis Louis Pasteur University, 1995.

[STE 97] STEWART G.W., *Afternotes goes to Graduate School – Lectures on Advanced Numerical Analysis*, SIAM, Philadelphia, 1997.

[STU 10] STUART A., "Inverse problems: a Bayesian perspective", *Acta Numerica*, vol. 19, pp. 451-559, 2010.

[SUN 94] SUN N.-Z., *Inverse Problems in Groundwater Modeling*, Kluwer, Dordrecht, 1994.

[SYM 93] SYMES W.W., "A differential semblance criterion for inversion of multioffset seismic reflection data", *Journal of Geophysical Research*, vol. 98, pp. 2061–2073, 1993.

[SYM 94] SYMES W.W., KERN M., "Inversion of reflection seismograms by differential semblance analysis: algorithm structure and synthetic examples", *Geophysical Prospecting*, vol. 42, pp. 565–614, 1994.

[TAR 00] TARANTOLA A., MOSEGAARD K., "Mathematical basis for physical inference", available at: www.ipgp.fr, 2000.

[TAR 05] TARANTOLA A., *Inverse Problem Theory and Model Parameter Estimation*, SIAM, Philadelphia, 2005.

[TOR 91] TORCZON V., "On the convergence of multidimensional search algorithms", *SIAM Journal on Optimization*, vol. 1, pp. 123-145, 1991.

[TRE 97] TREFETHEN L.N., BAU III D., *Numerical Linear Algebra*, SIAM, Philadelphia, 1997.

[VER 90] VERSTEEG R., GRAU G. (eds), *The Marmousi Experience*, European Association of Exploration Geophysicists, 1990.

[VIG 83] VIGOUROUX P., VANÇON J.P., DROGUE C., "Conception d'un modèle de propagation de pollution en nappe aquifère – Exemple d'application à la nappe du Rhin", *Journal of Hydrology*, vol. 64, nos. 1-4, pp. 267–279, 1983.

[VOG 02] VOGEL C.R., *Computational Methods for Inverse Problems*, SIAM, Philadelphia, 2002.

[WAH 77] WAHBA G., "Practical approximate solutions to linear operator equations when the data are noisy", *SIAM Journal on Numerical Analysis*, vol. 14, pp. 651–677, 1977.

Index

A, B, C

D

E

F, G

H, I

S

T, U, V, X

Other titles from

iSTE

in

Mathematics and Statistics

2015

DE SAPORTA Benoîte, DUFOUR François, ZHANG Huilong
Numerical Methods for Simulation and Optimization of Piecewise Deterministic Markov Processes

DEVOLDER Pierre, JANSSEN Jacques, MANCA Raimondo
Basic Stochastic Processes

LE GAT Yves
Recurrent Event Modeling Based on the Yule Process
(Mathematical Models and Methods in Reliability Set – Volume 2)

2014

COOKE Roger M., NIEBOER Daan, MISIEWICZ Jolanta
Fat-tailed Distributions: Data, Diagnostics and Dependence
(Mathematical Models and Methods in Reliability Set – Volume 1)

MACKEVIČIUS Vigirdas
Integral and Measure: From Rather Simple to Rather Complex

PASCHOS Vangelis Th
Combinatorial Optimization – 3-volume series – 2nd edition
Concepts of Combinatorial Optimization / Concepts and Fundamentals – volume 1
Paradigms of Combinatorial Optimization – volume 2
Applications of Combinatorial Optimization – volume 3

2013

COUALLIER Vincent, GERVILLE-RÉACHE Léo, HUBER Catherine, LIMNIOS Nikolaos, MESBAH Mounir
Statistical Models and Methods for Reliability and Survival Analysis

JANSSEN Jacques, MANCA Oronzio, MANCA Raimondo
Applied Diffusion Processes from Engineering to Finance

SERICOLA Bruno
Markov Chains: Theory, Algorithms and Applications

2012

BOSQ Denis
Mathematical Statistics and Stochastic Processes

CHRISTENSEN Karl Bang, KREINER Svend, MESBAH Mounir
Rasch Models in Health

DEVOLDER Pierre, JANSSEN Jacques, MANCA Raimondo
Stochastic Methods for Pension Funds

2011

MACKEVIČIUS Vigirdas
Introduction to Stochastic Analysis: Integrals and Differential Equations

MAHJOUB Ridha
Recent Progress in Combinatorial Optimization – ISCO2010

RAYNAUD Hervé, ARROW Kenneth
Managerial Logic

2010

BAGDONAVIČIUS Vilijandas, KRUOPIS Julius, NIKULIN Mikhail
Nonparametric Tests for Censored Data

BAGDONAVIČIUS Vilijandas, KRUOPIS Julius, NIKULIN Mikhail
Nonparametric Tests for Complete Data

IOSIFESCU Marius *et al.*
Introduction to Stochastic Models

VASSILIOU PCG
Discrete-time Asset Pricing Models in Applied Stochastic Finance

2008

ANISIMOV Vladimir
Switching Processes in Queuing Models

FICHE Georges, HÉBUTERNE Gérard
Mathematics for Engineers

HUBER Catherine, LIMNIOS Nikolaos *et al.*
Mathematical Methods in Survival Analysis, Reliability and Quality of Life

JANSSEN Jacques, MANCA Raimondo, VOLPE Ernesto
Mathematical Finance

2007

HARLAMOV Boris
Continuous Semi-Markov Processes

2006

CLERC Maurice
Particle Swarm Optimization

Printed and bound by CPI Group (UK) Ltd, Croydon, CR0 4YY

13/09/2026

14961024-0003